Re-thinking Mobility

Contemporary sociology

VINCENT KAUFMANN
École Nationale des Ponts et Chaussées, Paris
École Polytechnique Fédérale de Lausanne

ASHGATE

Published by
Ashgate Publishing Limited
Gower House
Croft Road
Aldershot
Hampshire GU11 3HR
England

Ashgate Publishing Company
Suite 420
101 Cherry Street
Burlington, VT 05401-4405
USA

Ashgate website: http://www.ashgate.com

British Library Cataloguing in Publication Data
Kaufmann, Vincent
  Re-thinking mobility : contemporary sociology. -
  (Transport and society)
  1.Transportation - Social aspects
  I.Title
  303.4'832

Library of Congress Control Number: 2002109903

ISBN 0 7546 1842 0

Printed and bound by Athenaeum Press, Ltd.,
Gateshead, Tyne & Wear.

Contents

Acknowledgements

I would like to thank Yves Ferrari, Michael Flamm, Robert Gibb, Jean-Marie Guidez, Christophe Jemelin, Dominique Joye, Fritz Sager, Martin Schuler and Catherine Smith with whom I conducted the various pieces of research presented in this work. I also wish to express my gratitude to John Urry, Bülent Diken and Jean-Marc Offner for the highly stimulating exchanges which we had during the writing of this book.

Introduction

Technical transportation and telecommunications systems are generating increasing numbers and forms of speed potentials. These systems, extensively used, have produced a new compression of time and space that has manifested itself in the robust growth of flux. Beyond the observation of this fact are the changes that are at the heart of a scientific debate about the impact of speed potentials on territory and on social structures. Although some describe this impact as the erosion of territory and social structures to the benefit of a *fluid* world that is more or less binary or separated into haves and have-nots, other researchers are much more prudent, preferring instead to refer to this occurrence as simply a change in the structure of spatial scales.

The present work proposes to consider the debate about the fluidification of society from the angle of the following problem: to what extent does the gradual elimination of certain factors of differentiation reflect the fact that social structure and territories are being created around new aspects whose definitions are largely absent from the tool-boxes of researchers?

This book represents a foray; it certainly does not claim to be an exhaustive study of the field. Its purpose is to make a contribution by exploring the implications of spatial mobility, which is at the centre of the debate on social fluidification. All too often, mobility is evoked as a preferred indicator to explain the compression of space-time and to describe its impact. However, in failing to distinguish speed potentials themselves clearly from their use, these analyses veer towards technological determinism, or else towards the normative domain. In order to avoid this trap and approach mobility as a possible new factor of social differentiation, the motivations underlying mobility must be explored.

In this context, this book is devoted to proposing the concept of *motility* and to testing its heuristic qualities. Motility regews to the system of mobility potential. At the individual level, it can be defined as the way in which an actor appropriates the field of possible action in the area of mobility, and uses it to develop personal projects. Using this concept, this work proposes to review the empirical data of four researchers on the relationship between transport systems and actors' mobility. This examination, which also provides an opportunity to study in greater depth the little-known field of the sociology of mobility, is carried out through the following general question: *to what extent can the speed potentials generated by technological transportation systems be considered as vectors of social change?*

The book is divided into eight chapters, the first two of which present a synthesis of the existing controversies surrounding social fluidification from the point of view of theoretical debates (Chapter 1) and of empirical research (Chapter 2). In response to the limitations revealed by this examination, which are the lack of dialectic between theory and empirical research, the confusion between speed potentials and spatial mobility, and the normative nature of many pieces of work, Chapter 3 proposes to rethink mobility using the concept of motility. Chapters 4 to 7 pursue this vein, presenting specific research results and highlighting the contributions of the concept of motility. These contributions are summarised in Chapter 8, which closes the book but opens new doors. The conclusion of the research work is that, although it does appear that society is adhering increasingly to the network model, this occurrence is not accompanied by social fluidification. Mobility, a central value of our culture, appears as an indicator of inequality; in particular, the analyses undertaken suggest that motility is becoming a type of capital in much the same way as education or social contacts. This last aspect invites us to change the tack of our approaches, and opens the door to a new form of general sociology.

1 Questioning Fluidity

Time-space compression and its implications

Technological advances in transport and telecommunications provide considerable potential for speed. Increases in speed have effectively nullified distances and allowed for immediacy in the circulation of information and ideas. The extensive use of these new speed potentials, which has contributed to their development, has produced a time-space compression, the social and spatial consequences of which are much debated. For many analysts such as David Harvey, this shrinkage of time-space questions the very basis of societies: '(...) we have been experiencing, these last two decades, an intense phase of time – space compression that had a disorienting and disruptive impact upon political-economic practices, the balance of class power, as well as upon cultural and social life' (Harvey 1990: 284). If one goes along with these authors, the compression of time-space could be likened to a sort of economic, social and cultural revolution we are all part of into whether we like it or not. Boden and Molotch sum up this interpretation well:

> The new communication and information technologies so heighten global and temporal integration that many regard them as harbingers of a deeper transformation in social relations. In the same way that previous commentators credited the steam engine and atomic bomb with ushering in new modes of civilisation, so the 'communication revolution' putatively brings a new form of existence. (...) People can live where they want without loss of friendships or kin solidarity, and capitalism can progress beyond fixed production lines and trading centres into a contemporaneous, multi-locational, non-stop world market (Boden and Molotch 1994: 257).

The existence of such upheavals is, however, controversial. Other authors, including Boden and Molotch themselves, are much more cautious with regard to the impact of time-space compression on social relations, and take the view that this aspect has been considerably exaggerated: '(...) we argue that the consequences for social life, whether benign or nefarious, have been exaggerated. The robust nature and enduring necessity of traditional human communication procedures have been underappreciated' (Boden and Molotch 1994: 258). From these very global considerations about the world and its future is born a question that has haunted sociology for ten years: are we really witnessing a fluidification of society, driven by the growth of mobility and the movement of goods, information and ideas?

The social fluidity question is nothing new. It harks back to the dream of a classless society guaranteeing equal chances for all. With the advent of modern systems of communication and transport and their impact, however, the argument takes on another twist. According to classical sociology, a fluid society presents no barriers and permits the individual to move vertically in socio-professional space on a strictly meritocratic basis. Sociology has been grappling with this question since the 1920s and the works of Pitirim Sorokin. The idea has been developed considerably since the 1960s, notably with the work of Lipset, Zetterberg and Bendix and that of Blau and Duncan (see Cuin 1983). In this often ideologically oriented bevy of work, fluidity has very positive connotations as a conveyor of advancement in social justice. Compared to this traditional sociological perspective, the fluidity debate that currently concerns the social sciences differs in at least three ways. First of all, it concerns horizontal movement as much as vertical movement in social space. In this sense it abandons the idea of a single aspirational model – socio-professional success – which individuals must pursue. Secondly, it includes transport and communication systems as actants or manipulators of time and space. Although in the classic works on social mobility, space and time did occupy a permanent but often reified position, they now occupy an unambiguously central position in the debate. Thirdly, this debate is no longer restricted to the domain of work, but is more global, concerning itself with different spheres of activity and their relationship with time and space. In short, the fluidification debate is much more far-reaching than the simple question of transferring from one social category to another. It concerns all barriers, constraints and margins for manoeuvre that individuals are confronted with throughout their lives and can be summed up in the following question: *Does the compression of time-space expand the room for manoeuvre that each individual has available to him/her in the course of his or her life?*

The 'new' debate surrounding fluidity has already caused considerable amounts of ink to flow in social science, fuelling controversies over technical determinism and the impact on society of expanded exchange of goods, information and mobility of people. The debate more or less revolves around four theoretical standpoints: the areolar model, the network model, the liquid model and the rhizomatic model. These correspond to specific angles of analysis and are not totally antinomic. Each one suppports a different view of the relationship between mobility and fixity.

The areolar model: the predominance of fixity

The areolar model postulates substantial determination of movement by the socio-spatial structure. It can be defined as areolar insofar as it refers to regions delineated by borders. This perspective has long represented mainstream sociology, permeating different strands of thought from Marxism to certain variants of methodological individualism. In this first model, space-time is conceived as a 'limited form' or a static space that is defined by borders (Montulet 1998), which de facto neglect space as relevant category for sociology (Tickamyer 2000). The whole of society is structured around 'limited spaces' such as social classes or states that make up a system and define a functional and cultural unity that establishes the unity of the individual. Mobility in this perspective consists of moving from one social or spatial category (or both) to another. There is thus an orientation to this movement: it has an origin and a destination. But its medium is considered neutral. Technological systems are created by societies and are not, therefore, vectors of change in themselves.

A large proportion of the conceptual and methodological apparatus of sociology is based on this model, which implies that all social groupings belong to clearly defined categories (Wellman and Richardson 1987). The majority of the available statistical sources implicitly refer to it, their criteria of social (socio-professional and household composition) and spatial (countries and administrative regions) differentiation are based on spaces predefined as relevant, homogenous and defined by borders.

The areolar model is the dominant paradigm of stratification analysis. One of the most typical applications of this model is undoubtedly the study of another classic area of sociology, social mobility. And one of the characteristics of this field is the predominance of quantitative approaches based on the analysis of intergenerational mobility tables. These mobility tables are constructed on a national scale, sometimes differentiating the sexes and frequently comparing nations. In so doing, they measure the movements between origins and destinations, whilst presupposing that the different social classes have the same homogeneity and that they can be compared in time and space, which is also considered stable.

The areolar model is often criticised for being static (Touraine 1995, Urry 2000a, Beck 1992, Ascher 2000). The most fundamental objection levelled at it is that it is based on the notion of society as corresponding to a natural organism which no longer exists (or never actually did) for want of unity between cultural identities, political action and economic power (Dubet 1994: 14-15). It is consequently wrong and illustrative of how research can become enmeshed in pre-stablished and closed categories. For its detractors, the proof of its obsolescence is provided by the decline in the differentiating power of a good many of these analytical categories.

The network model: the mobility-fixity dialectic

The reticular model affirms the predominance of structures formed around networks. In this perspective, recently formalised by Manuel Castells, social structures are controlled by elites who move in spaces of flux: 'Elites are cosmopolitan, people are local. The space of power and wealth is projected throughout the world, while people's life and experience is rooted in places, in their culture, in their history' (Castells 1996: 415-416). In this view, analycal categories from classical sociology continue to be relevant for the understanding of social structure. Social stature, therefore, depends on both economic and cultural capital (based on education and socio-professional status) and on social capital, understood as 'social networks'. The dynamic dimension of the network model is assured by this social capital, which becomes progressively freed from contiguity to become connex (Offner and Pumain 1996), thanks to the opportunities offered by rapid transport and advanced means of communication. In this process the social structure becomes both fixed and mobile.

Access to information is a key element in this model and the medium of communication also at the heart of the analysis. Whilst in the areolar model transport and means of communication are merely neutral devices in a social and/or spatial mobility between an origin and a destination, in the network model they are likely to favour a form of social and/or spatial mobility by means of the access they permit (Lemieux 1999: 3).

The branch of sociology concerned with social networks has significant parallels in sociology of the family (Parsons 1960, Girard 1964), urban research (Sassen 1991, Veltz 1996), urban politics (Judge et al 1995) and economic sociology (Latouche 1998). It is even tending to assert itself as a paradigm – structural analysis – in response to the limitations of the areolar model. It has also formed the basis of conceptual developments around the notion of network (see, for example, the work of the Groupe de recherche réseaux du CNRS – Offner and Pumain 1996) or the actor-network (see, for example, Law and Hassard 1999). The sociology of social networks has also been at the forefront of methodological advances (Degenne and Forsé 1994).

The liquid model: the predominance of mobility

From this third point of view, speed brings about the progressive weakening of the social structure and of its categories in favour of a world organised around mobility. This society can be qualified as liquid (to borrow the term from Bauman 2000, and in contrast to the solid society of the areolar model) insofar as it takes the form of its habitat and is thus fundamentally ambivalent, heterogeneous and reversible. It is the era of ephemerality and consumerism, the throw-away society. In this context, human beings are ever-changing. One can just as easily throw out one's material waste as one's values, lifestyles or attachment to people and places (Montulet 1998: 126). Social analysis thus becomes dual: on one side there are the dominant groups who evolve in a mobile space, and on the other are the excluded who evolve in a fixed space, chained to localism through circumstances, but without there actually being any need to attach themselves to it. 'People who move and act faster, who come nearest to the momentariness of movement, are now the people who rule. And it is the people who cannot move as quickly, and move conspiciously yet the category of people who cannot at will leave their place at all, who are ruled' (Bauman 2000: 119-120). Fast transport and new information technologies are at the heart of these evolutions in which they are actors in the role of 'speed creators'.

This highly individualised world is characterised by the end of 'mutual engagement': 'The end of the era of mutual engagement: bctween the supervisors and the supervised, capital and labour, leaders and their followers, armies at war. The prime technique of power is now escape, slippage, elision and avoidance, the effective rejection of any territorial confinement with its cumbersome corollaries of order-building, order-maintenance and the responsibility for the consequences of it all as well as of the necessity to bear their costs' (Bauman 2000: 11). Collective sensibility, which anchored itself on macro-social entities, is replaced by sociabilities functioning around free choice, leading to neo-tribalism (Frétigné 1999: 86). This absence of mutual engagement brings to light the fundamental ambivalence of mobility among those who travel for business, career or pleasure purposes, and those who are obliged to do so for survival (Bauman 2001: 38). In the first scenario it is the expression of the reversibility of the throw-away society. In the second it is actually the expression of the constraints that the people excluded from the fluid world are subjected to.

The development of the liquid model relied heavily on the sociology of consumerism and lifestyles (Shields 1992) as well as on research concerning tourism (Urry 1999). On a conceptual level, it is evident that the notion of social exclusion is derived from the liquid model (Touraine 1992). This notion is based on the observation of a vertically stratified society sliding down to a horizontally stratified society that is exclusive and dualist. 'Modern society had a vertical base; class relations prevailed; one was either at the top or the bottom. Today a slide towards a horizontal society is occurring, where individuals are to be found either at the centre or on the periphery' (Frétigné 1999: 88) (my translation).

The rhizomatic model: the disappearance of mobility and fixity

The rhizomatic model posits the advent of a world dominated by technology. It takes its inspiration from many sources, ranging from the work of Jacques Ellul (1954) on the subject of technology, to that of Gilles Deleuze and Felix Guattari about deterritorialisation (1987). In this perspective, time-space compression is such that it brings with it the disappearance of 'limited forms' to the benefit of an 'organizing form concept where the populating of time supplants the populating of space' (Couclelis 1996). Society is like a body without organs. Space is thus undefined and open. It is a set of opportunities in perpetual reorganisation, a rhizome. The world is now nothing but a vast interface. 'The instantaneity of ubiquity results in the atopia of a simple interface. After distances of space and time, speed distance abolishes the notion of physical dimension' (Virilio 1984: 19) (my translation). In this view of things modernity and postmodernity are smothered in a non-stratified world defined by individual desire, without spatial or temporal differentiation.

In the rhizomatic model this disappearance is prompted by the immediacy enabled by information and communication technologies. The compression of space-time here is complete and its social impact is radical. The notion of mobility no longer has any sense, as the crossing of borders is no longer possible, space being no more than a rhizome. The individuals in this case are reduced to living in a largely immaterial world, transporting their territories with them.

Largely confined to post-structuralist philosophy and to certain catastrophist futurologies espousing an exaggerated technological determinism (Töpfler 1980), this model has not given rise to the development of a paradigm in social sciences. Vehemently criticised as a global theory (Lévy 1999; Ascher 1998), it is, however, referred to in very diverse fields. For example, the idea of a rhizome is currently used in the field of migration to illustrate constructions of identity not supported by space. This model has especially been used in the field of gay and lesbian studies (Fortier 2000).

Two ideal types

Each of these four models can be reduced to two ideal types about the intelligibility of space and time: *the world of sedentary space and the world of nomadic space.* To use the words of Deleuze and Guattari again, 'sedentary space is "striated" by walls and paths between the fences, whereas nomadic space is smooth, marked only by lines which are effaced and displaced in the course of the journey' (Deleuze and Guattari 1987). These two worlds can be illustrated by comparing the games of Chess and Go:

> Chess pieces are coded; they have an internal nature and intrinsic properties from which their movements, situations and confrontations derive. They have qualities; a knight remains a knight, a pawn a pawn, a bishop a bishop. (…) Go pieces, in contrast, are pellets, disks, simple arithmetic units, and have only an anonymous, collective, or third-person function. (…) Go pieces are elements of a non-subjectified machine assemblage with no intrinsic properties, only situational ones. (…) Chess is indeed a war, but an institutionalized, regulated, coded war, with a front, a rear, battles. But what is proper to Go is war without battle lines, with neither confrontation nor retreat, without battle lines even: pure strategy,

whereas chess is a semiology. Finally, the space is not at all the same: in chess, it is a question of arranging a closed space for oneself, thus of going from one point to another, of occupying the maximum number of squares with the minimum number of pieces. In Go, it is a question of arraying oneself in an open space, of holding space, of maintaining the possibility of springing up at any point: the movement is not from one point to another, but becomes perpetual, without aim or destination, without departure or arrival. The 'smooth' space of Go, as against the 'striated' space of chess (Deleuze and Guattari 1987: 352-353).

In each of these two worlds, movement takes on a specific meaning. In sedentary space with its confined structures it can be understood as the crossing of territorial boundaries formed by social and spatial stratification. In nomad, non-structured space, movement is by definition 'immobile', insofar as the nomad moves with his or her territory. He or she has no origin or destination and is unable to cross any social or spatial borders/frontiers.

The four models take up positions in relation to these two ideal types, either by rejecting one of the terms (the areolar model and the rhizomatic model) or by combining them (the network and liquid models). In this sense, each model can be situated on an axis ranging from the world of sedentary space (structured, confined) to a world of nomadic space (non-structured, infinite).

Table 1.1   Sedentary space – nomadic space axis

Structured confined				Non-structured infinite
Areolar	Network	Liquid	Rhizomatic	

Four interpretations of fluidification

This brief introductory discussion has inevitably been schematic, given that the aim of this book is not to carry out an in-depth analysis of the history of these different models and the various currents which cross them. It could be completed in many ways, in particular by going deeper into the links between the different positions and trends of thought that have influenced the social sciences during the last thirty years. Despite the limitations, it does enable a synthesis of the debate surrounding fluidification. In light of this introduction it is effectively possible to represent fluidification as a narrower or wider movement, from the structured and confined pole to the non-structured and infinite pole. So, each of the four models just presented represents an interpretation of this movement that defines different key issues for sociological research.

- From the point of view of the areolar model, *this movement does not actually take place*; fluidification is no more than an illusion linked to changes in spatial-temporal scales which have no real impact on structures. In effect it is the reification of space that is the basis of the illusion of fluidification. Pools of workers widen and consequently these pools no longer correspond to administrative areas, and it is therefore necessary to find new relevant differentiating scales (Joye et al 1995, Bassand 2000). Time-space compression is here considered as a phenomenon with no real impact on societies. In support of this theory, one can refer to the relative and unequal character of the compression of space-time, which would do no more than reproduce social and spatial structures under another form without changing them drastically either on the level of centrality hierarchy or on the level of social inequalities.

- From the network model viewpoint, *movement* is the fact of the *connexity of the networks of actors who break the spatial-temporal unity of societies*. We are experiencing a passage from the areolar model to the reticular model. Fluidification is first and foremost time-space-related and comes from the breaking down of contiguity. Social capital increases in importance in the construction of social positions, notably due to the fact that access to information becomes a key issue (Castells 1996, Putnam 2000). The social aspect remains stratified and organised around societies defined by state frontiers. But the importance of networks of actors gives these societies a dynamic character. Modern systems of communication and transport are at the heart of change in that they allow the speed of circulation which makes connexity possible. The processes of globalisation can be cited, in support of this theory.

- The liquid model posits *a movement of social and spatial fluidification by the reversal of the balance between fixity and mobility, to the benefit of the latter.* This is a reversal that marks the end of modernity. From this point of view, spatial and social mobility are indistinguishable. We are moving from a world organised around sedentarity to a world organised around mobility. In this third view, the social aspect is characterised by a radical fluidification with the passage from the areolar model to the liquid model. The result of this is an obliteration of social and spatial structures inherited from modernity, creating in its place an inclusive/exclusive binary stratification, where those included have endless opportunity to escape, rendering all experience reversible. The compression of space-time is central to the construction of the liquid model (Bauman 2000, Tourraine 1992). In support of this theory, one may cite the example of the decline in explicative power of areolar social categorisations and the importance of lifestyles and consumer practices.

• In the rhizomatic perspective, *movement is total and is associated with a generalised ubiquity*. Total fluidification brings with it the disappearance of social and spatial structures and the territories that are associated with them. There is no longer any movement possible as there are no longer any borders to cross. The passage from the areolar to the rhizomatic model happens in stages and the reticular and liquid models can be understood as intermediary stages. The disappearance of the world of sedentary space is a process that will emerge as the rhizomatic world due to the compression of space-time procured by new information and transport technology. This total fluidification is determined by technologies which fashion the world in their image, causing a disappearance of the real world in favour of a virtual world (Virilio 1984, Albertsen and Diken 1999). In support of this theory, the ubiquity and virtualisation of capital as well as the immediacy of information transmission of and its impact on the financial world can be cited.

Reconsidering the debate

Fluidification is one of the ultimate sociological questions. It raises the issue of the role of space and modern technology in the construction of social phenomena and the future of social structures and power in the contemporary world. However, central as it is, the debate surrounding fluidity has several limitations, making a discussion of the approaches just presented necessary. Three aspects that are particularly worrying are the absence of a dialectical relationship between theory and empirical research; the confusion between mobility and potentiality for movement; and the formulation of very generalised positions. These will now be dealt with in turn.

The absence of dialectical theory – empirical research

In research work relating to fluidification, empirical observation often occupies a paradoxical position. Although it is used as the ultimate proof of the veracity of the different models just presented, this use tends to be selective and partial and the measuring instruments imprecise. In fact, the dialectic between theory and empirical research is often absent from the models just introduced. The reading of a number of works on this theme arouses suspicion. One could easily get the impression that the authors use the results of empirical research to justify pre-established theoretical positions rather than to discuss them (Kaplan 1996). Any results that would contradict the established positions are eliminated. The use of Marc Augé's 'Non-Places' is completely symptomatic of this tendency. Ritually cited as proof of the disappearance of territories, this work in effect contraposes the idea of places as social, historical and identity references with non-places, which are defined as non-social, non-historical and non-identity, which cannot be related to and are non-historic (Augé 1992: 100). Augé defines the traveller's space as the archetypal non-place (Augé 1992: 110). Criticism of this book, of which there has been much since it was published in 1992, is systematically ignored by those who cite it, especially that criticism which relates to the fact that places of mobility can perfectly well be references in relational terms, and even in terms of identity, through memory especially (Chalas 1997, Lévy 2000).

The questioning which forms the basis of the fluidification debate is of the same nature as this criticism. In its general form, which is usually based on a question along the lines of: 'Are we experiencing a modification of the balance between mobility and fixity driven by the mobility of people and the circulation of goods, information and ideas?', three aspects considered as established facts merit an empirical discussion:

Firstly, it is not certain whether spatial mobility in general *is* increasing. Although the average distances covered during the course of daily life are increasing, the number of journeys per person per day has remained stable (since recording began in the 1950s). This figure is between 3.5 and 4 (Salomon et al 1993). Pedestrian mobility is in constant decline. It may be undeniable that international tourism is strongly increasing, but on the other hand the coastal resorts of the north in several countries have become more and more deserted since the 1960s. Examples of this are Blackpool and Morecambe in England. In Morecambe, for example, the number of hotels decreased from 640 to 267 between 1973 and 1987 (Urry 1990: 31). In the same way, local mobility is losing ground to new forms of mobility like long distance commuting or travel to and from a second home (Ascher 1998). Likewise, although it is undeniable that we are witnessing a growth of transmitted information, if this is only increasing in terms of volume, one may legitimately question whether this growth is accompanied by a growth in the reception of this information. Is it possible that excessive information destroys real information? It is undeniable that email has burgeoned at the expense of the letter, but its content is not of the same nature. It is necessary to agree on what one means by the growth of mobility and the circulation of information. If it is a question of increasing distance and the speed of flux, the affirmation is correct. But if it is a case of the number of journeys and the receiving of information, it is as well to remain cautious. More than growth, it seems that we are dealing with substitution phenomena between forms of mobility and circulation.

Secondly, the network, liquid and rhizomatic models each establish in their own way a parallel between fluidity in space-time and social fluidity, although these are two orders of reality which do not necessarily go together. Just because one observes an increase in spatial impact speed flux does not necessarily mean that this is consistent with a growth in social fluidity. In particular, movement in geographical space can quite easily be seen as a constraint on and not a widening of possibilities to move in social space. Two-career households provide a very good example of this type of constraint. When the two working individuals have jobs in different urban centres, and, for example, they begin to co-habit, there will always be choices to be made. One of the two may eventually give up his or her job, but generally choices involve a combination of daily commuting and deciding on residential location (Bonvalet and Brun 1998).

Thirdly, the fluidification issue suggests that the impact of the growth of flux on social aspects of life always leads towards a fluidification and a disappearance of territories. But what if this expansion were to create a hardening of certain structures and certain territories? Beyond the dichotomous analysis of Zygmunt Bauman (and numerous works about social exclusion), which demonstrate well how the fluidity of some assigns others to sedentarity, such a hardening may take other forms. Flux does in fact reinforce certain territories. For example, the development of mass media may reinforce the territorialisation of languages. Switzerland, multicultural state par excellence, is absolutely revealing of this phenomenon. The progressive construction of media spaces corresponding to the territorialisation of languages reinforces cultural cleavages between speech communities (Du Bois 1983). Also in the area of languages, one can cite the impact of moving house on the hardening of the linguistic border between Walloons and Flemish in the region of Brussels. The growth of the francophone population in the Flemish-speaking territory has provoked the adoption of a very strict legislative arsenal protecting the territorialisation of languages in the Flemish region.

These three aspects illustrate the necessity of coming out of the world of theory to develop a dialectic between theory and empirical research.

The confusion between movement and potentiality of movement

Arguments relating to fluidification are largely based on the notions of flux, movement and mobility. These ideas refer to movements that are actually carried out. In this domain, as in many others in the social sciences, the observation of a tendency is insufficient to explain the social. The most convincing example of this is undoubtedly intergenerational social mobility. In the majority of western countries an increase in intergenerational social mobility has been observed. An increasing section of the population occupies a different socio-professional category from that of their mother or father. But it is by no means certain this it a sign of social fluidification. This question can be related by analogy to the old distinction between structural social mobility and net social mobility. Just because a mobility flux can be observed, one cannot necessarily attribute it to a free circulation of individuals; it could be purely a reflection of structural changes. The confusion between an observation and the underlying rationales is not only frequent in research papers; it can also be found more generally. When the French government decided on the target of 80% of young people getting through the 'Baccalaureat' for the sake of the democratisation of education[1] it confused equality of opportunity and volume of diplomas issued, and thus contributed to a knock-on effect in education. More diplomas are issued so they are effectively worth less. Young people now have to study for longer and longer to reach the same social position as their parents.

To return to the area that concerns us, confusion between mobility and potential for mobility leads to a form of technological determinism. Mobility does not necessarily increase just because there is a greater potential for mobility thanks to faster transport and new information and communication technology, as supporters of the network and liquid models and especially the rhizomatic model suggest. It is important to differentiate properly between these two types of phenomena. The fact of being able to get from Paris to Brussels on the TGV in an hour and a half or from London to any European capital in two hours does not necessarily mean that all the inhabitants of these cities are going to carry out such journeys (they must have a reason for going there and the financial means to do so!). Having Internet access at home potentially allows access to a universe of information and opens up all sorts of possibilities. However, this potential is not necessarily used. We know that even when it is, it is done very selectively (Rallet 2000). The crux of the debate over social fluidification is whether or not the compression of time-space goes hand in hand with a decrease in certain social constraints that discourage action. It is thus a question of analysing who has access to which relevant technology and the degree of freedom afforded by the usage of this technology.

To regain an analytical stance that differentiates potential actual practices and their impact is indispensable to be able to debate possible fluidification. The question is what to do with these standpoints that claim that the very existence of technological systems such as new information and communication technology leads to a radical virtualisation of the world.

Assertions that hinder discussion

The different theoretical standpoints relating to fluidification are too often exclusive and generalised. Explicitly linked to schools of thought such as structural-functionalism, post-modernism, post-structuralism, etc. to a greater or lesser degree, they often seem to be unaware of each other. This lack of awareness is a result of the fact that they evolve in different fields of research which, in the opinion of François Dubet – and I agree with him on this point –

causes three major disadvantages: 'The first is that general theories are in fact treated as middle range theories. The second is that intellectual fashions play a central part owing to the weakness of the lack of choice. The third disadvantage may be the abandonment, with little consideration, of the ambitions of classic sociology' (Dubet 1994: 14) (my translation). When indifference ends, defamation often replaces it. The words of Bourdieu on the sociology of Bauman or the demolition of the work of Virilio carried out by Pierre Lévy are absolutely symptomatic of this situation[2]. The 'wars of the religions' are undoubtedly part of sociology's charm but they are also an obstacle to the advancement of knowledge. The situation is paradoxical: the four models presented are often expressed in a generalized form but used in a specific way. When there is discussion between them it is in the form of a lampoon. A critical analysis of these four models shows that it could not really be any other way. The reasons are as follows:

Firstly, the very globalised reflections proposed by the four models often seem to take the form of ideology, which makes debate impossible. Certain formulations of the network concept, for example, serve as a vehicle for the myth of progress and sink into a form of technological determinism (Offner 2000, Claisse and Duchier 1993). This type of theory is, in fact, totally in harmony with neo-liberal universalist ideology which promotes the world of nomad space over the world of sedentary space in the name of the necessity to fight against borders, considering these archaic. As John Urry sums this up, 'Those with economic interests in promoting capitalism throughout the world argue that globalisation is inevitable and that national governments should not intervene to regulate the global market-place' (Urry 2000a: 12). This neo-colonialist ideology does not espouse physical occupation of territories, but control of nomadic space. It draws its legitimacy from the notion of progress. It is nevertheless difficult not to see it as a strategy to dominate the world economy. An example among hundreds of others is the battle of the 'appellations contrôlées' for wines (AOC). This battle is extremely significant from this point of view. The United States challenge such regional sedentary designations, in favour of designating wines in the nomadic terms of vine types (cépages) (Ascher 2000).

Secondly, the expression of these stances is too often rather sensationalist. Thus, the messages of certain proponents of the liquid and rhizomic models are often tinged with technological enthusiasm or alternatively with an alarmism predicting a return to obscurantism. As Urry puts it, 'There are global enthusiasts who see these processes as producing a new epoch, a golden age of cosmopolitan "borderlessness". This epoch offers huge new opportunities, especially to overcome the limitations and restrictions that societies and especially nation states have exercised on the freedom of corporations and individuals to treat the world as "their oyster". Others describe globalisation not as a borderless utopia but as a new dystopia. The global world is seen as a new medievalism, as the "west" returns to the pre-modern era. (…)' (Urry 2000a: 13). The question which follows from this is: How does one return to a real debate in such an intense atmosphere? The evolutionist zeal contained in certain writings expresses an obsession akin to sensationalism. Commercial logic and fashions have heavily influenced the social sciences for a long time now, sometimes creating situations where each publication tries to outdo the last, with the final aim being more one of recognition for the author than the furthering of knowledge. This recognition is vouched for by the number of citations of the 'product'. Unfortunately, the fluidification question has not escaped this phenomenon. We are, thus, bombarded by both optimistic and pessimsitic prophecies with a pronounced taste for scenarios predicting 'the end of' more or less everything.

Thirdly, the debate between the models is complicated by cultural issues. Several authors mentioned have developed their works in specific national contexts. The way that fluidification has been approached is thus based around different research traditions. Ulrich Beck's concept of *Risokogesellschaft*, for example, is born of the nuclear debate in Germany, a debate which has not had the same importance in numerous other European countries. Not taking these differences into account can cause misunderstandings which prevent the discussion from being established on firm ground. One of the significant examples of this point of view is undoubtedly the translation into English of the works 'Anti-Oedipus' and 'A Thousand Plateaus' by Deleuze and Guattari. This translation has generated a plethora of references in the United States concerning 'deterritorialisation' and 'nomadic subjects' (Kaplan 1996: 92). However, when one speaks of deterritorialisation in France, the connotations are very different from those connected to the same terms in the United States. In particular, the subversive character of the notion is lost. The two books by Deleuze and Guattari are a criticism of French culture in general, and of Republican Jacobinism and especially the great State institutions, something that Deleuze builds on in a collection of interviews (Deleuze et Parnet 1996: 47-53). Not to remember this or to take these texts at face value leads to misunderstandings that explain Caren Kaplan's position when she sees nothing more than a form of neo-colonialism deterrritorialisation/reterritorialisation (Kaplan 1996: 92), whereas, in reality, these texts are really above all about France 'intra-muros'.

From non-debate to dialogue between models

One point has become clear in the course of this critical examination; the fluidification debate is not really just one single debate. Although expressed as global theories, the four models discussed stand out as partial constructions of social reality. Nobody denies that certain areas of the economy work in a deterritorialising way, but does this mean that territories have actually disappeared? Nobody denies that the same consumer goods can be found throughout the whole world, but surely it is deceptive to see in the usage of the same objects the development of the same habits (Ascher 2001: 148) Similarly, nobody would deny that the classic dimensions of social stuctures have less power to explain behaviour than in the past. But is this necessarily a sign of the obliteration of social categories? Because each model is partial, when the four models are combined, as in reality, a more global vision of social reality can be achieved. There has not been any fruitful dialogue between the models, which have been treated separately in theoretical terms. They are expressed globally, sometimes in an ideological or normative manner, or conceived as consumer products. A true debate challenges them by demonstrating, to a greater or lesser degree, their partiality. But none of the advocates of the different models takes this risk, as each model corresponds de facto to specific research areas. The debate will thus have to come from the outside.

In order to overcome this paralysis, a mobile form of sociology is called for (Diken 1998). Not only does it need the flexiblility to move between different theoretical approaches and between empirical research of different types, but also between languages. Having a true discussion about fluidification also implies going beyond the myths of determinist relationships between networks and society and between networks and territories (Offner 2000).

In order to develop such an ambition, it is still necessary to base it on a paradigm and, above all, on a conceptual apparatus. It is precisely this last aspect which is lacking. On examining the matter more closely one realises that the gulf between theoretical debate and empirical research is a reflection of the absence of adequate conceptual tools for dealing with the question of mobility and flux. Moreover, this absence of tools is linked with the hegemony of the areolar model in numerous fields of sociology up until recently. This model considers movement in the geographical space of places to be irrelevant as an object of sociological research. Approaching the social aspects from the angle of potentialities also demands a conceptual apparatus which the mere notion of risk does not allow to cover. To develop a discussion between the four models involves having a conceptual apparatus available that is not based on one of the models but that nevertheless allows them to be tested. In the last analysis, *the limitations of the fluidification debate come down to a conceptual vacuum.* Conceptualising fixity does not seem to pose much of a problem. Conceptualising mobility, on the other hand, is more problematic (Dupuy 1991, Diken 1998).

Establishing a mobility paradigm

In order to tackle social fluidification, it is necessary to start from the basis of a neutral paradigm from the point of view of the four models introduced.

For the same reason it is essential not to base debate on an a priori view of society, which would stifle the debate before it had begun by attaching it to either the areolar model or the network model. I therefore propose to get rid of the very concept of society in order to replace it with an approach based on movement. By adopting such a standpoint I subscribe to John Urry's perspective (Urry 2000a) which proposes the replacement of the classic approach to society with one based on mobility and fixity. Such an approach would undoubtedly seem like sacrilege to numerous sociologists, the discipline having been built around the concept of society. However, this approach does not imply either the assumption of the disappearance of societies or their dilution. It is simply an intellectual stance. From the point of view I have taken, the existence of societies, whether they correspond to state borders or not, is no longer a presupposition, but a result.

The problem posed by such a change of position is not so much its expression as the absence of conceptual tools with which to study it. As I have already mentioned, the conceptual apparatus of sociology is effectively largely based on concepts derived from the areolar model, namely the clear delimitations of enclosed territories (Wellman and Richardson 1987, Montulet 1998). These concepts refer to static territories; you are either on the inside or the outside. But many of these clear distinctions are no longer such strong factors of social differentiation as in the past. This fact prompts certain authors, like Ulrich Beck, to say that they are no more than zombie categories: 'zombie categories' and 'zombie institutions' which are 'dead and still alive'. He [Beck] names the family, class and neighbourhood as the foremost examples of this new phenomenon (Beck, quoted by Bauman 2000: 6).

The problem of the lack of a conceptual apparatus therefore goes further than the sole question of fluidification. The example of urban sociology demonstrates the existence of this problem in all areas. Urban sociology has abandoned the notion of cities as areolar space contrasting with the countryside to replace it with an approach based on the urban notion that considers towns and cities as a reality going beyond morphological materialisation (Ascher 1995, Remy and Voyé 1992). Nevertheless, on a micro scale, this new approach is still, despite everything, limited on an empirical level by concepts based on the areolar model. Thus, the notion of neighbourhood, which many researchers have converged in considering an unsuitable analytical concept to describe actual social practices, remains a very commonly used concept. In the same way, social segregation, which has traditionally been formalised as areolar on the community or neighbourhood scale, is considered by many as very largely insufficient to capture the extent of the distribution of inequalities in urban space (Grafmeyer 1994, Roch 1998). This does not, however, prevent numerous authors from continuing to develop segregation indices.

Dealing with fluidification requires a rethinking of the concept of mobility in order to have a notion available that will allow microsociological as well as macrosociological research and to highlight differentiation factors. More specifically, it is a question of developing a precise instrument that can cross over between all four models presented and differentiate between movement and potential movement. To do this, a presupposition will serve as a starting point: *that copresence is a social constraint.* As Boden and Molotch put it: 'Copresent interaction remains, just as Georg Simmel long ago observed, the fundamental mode of human intercourse and socialization, a "primordial site for sociality" in Emanuel Schegloff's phrase. Modernity is made possible not by the substitution of new technologies for copresence but by a tensely adjusted distribution of copresence and the more impersonal forms across individuals, tasks, places, and moments' (Boden and Molotch 1994: 258).

John Urry clarifies this analysis by pointing out that the bases for copresence can be categorised into six types: 'Legal, economic and familial obligations either to specific persons or generic types of people: to have to go to work, to have to attend a family event, to have to meet a legal obligation, to have to visit a public institution; social obligations: to see specific people "face-to-face", to note their body language, to hear what they say, to meet their demands, to sense people directly, to develop extended relations of trust with others, to converse as a side-effect of other obligations; time obligations: to spend moments of quality with family or partner or lover or friends; place obligations: to sense a place or kind of place directly, such as walking within a city, visiting a specific building, being "by the seaside", climbing a mountain, strolling along a valley bottom; live obligations: to experience a particular "live" and not a "mediated" event (political event, concert, theatre, match, celebration, film [rather than video]); object obligations: to sign contracts or to work on or to see various objects, technologies or texts that have a specific physical location' (Urry 2000b: 12-13).

To accept copresence as the fundamental mode of socialisation amounts to considering the compression of space-time as largely the effect of modern transport and communication systems. Means of transport and communication procure not only speed but also increase the means of being mobile: more modes of transport, more destinations and more itineraries are now available. Access to these different possibilities and the way they are used by actors now become potentially important factors of differentiation and social distinction. It is precisely this argument that is defended by Prato and Trivero (1985). They describe mobility as having become a primary activity of existence and a vector of social status.

The concept of mobility that the fluidification debate needs thus refers to spatial mobility which enables the question at the basis of fluidification to be clarified as follows: *is the rapidity procured by technological transport and communication systems significant enough to allow an increase in the margins of manoeuvre in its controlling of individual life?*

Notes

1   The rate has gone from 26% in 1975 to 31% in 1980; 44% in 1990; 58% in 1995; 55% in 1999. Source: *Sciences Humaines*, no. 111, p. 28.

2   Due to the esthetic quality of the text, I couldn't resist reproducing two extracts from Pierre Lévy's satirical lampoon 'La pensée crash de Paul Virilio' (1999) about Virilio's 'La bombe informatique' when it appeared in 1998: 'Content merges with form in "la bombe informatique". The urgency, like the incoherence and the expression, and the thought are those of panic. This text expresses the mental flow of a person in the throes of panic who wants to frighten, even terrorise. This is not thought but extreme fear expressed in words. We were already used to the trash theory. Virilio innovates by inventing the crash theory. It is a sort of David Cronenberg of theory, a man with an accident fixation who has nothing more to offer than confused and incomprehensible scenarios, like those of a disaster movie. Out of thousands of other examples one moves directly from stereoscopic virtual reality helmets to the complete collapse of all culture: all this in brief, accelerated and at maximum theoretical speed. The intellectual wall of sound is broken.'

'Like the vicious circles into which whole sections of contemporary art are sucked with its painters who, not knowing how to paint, merely spit on the painting, here are the moving sands of theory, with its thinkers who don't know how to think and spit instead. The one who can spit fastest wins. Virilio announces the end of everything and succeeds in speeding it up, at least for the cultural sphere who take him seriously' (my translation).

2   Insights from Empirical Research

Introduction

Before developing a set of conceptual tools capable of dealing with mobility, which will be the aim of the following chapter, I will first re-examine the evolution of mobilities with reference to empirical research. In the previous chapter I presented four theoretical models corresponding to specific standpoints in relation to the fluidification of the social. In so doing I regretted the weakness of their links with empirical research. The question is thus, on the basis of research findings, what can be said about the development of mobility? What is known about it?

By way of approach to mobility research, I will adopt the phenomenological postulate that assumes the equality of the notions of identity, space and time, and deal with them together (Tarrius 2000: 38; Rapport and Dawson 1998: 3). This postulate is based on the idea of a paradigm of mobility (Tarrius 2000) that does not dissociate time, space and identity. Thus, I will assume that all mobility has repercussions on identity and, inversely, that an identity is built on mobilities. The advantage of this postulate is the fact of not making reference to any of the four theoretical models discussed in the previous chapter and allowing cross-overs between different disciplinary approaches. I want to distance myself from the approach to mobility that differentiates migrations, residential mobility, travel and daily mobility in all its forms.

By basing my analysis on this approach it is possible to describe the evolution of mobilities in terms of the three processes briefly touched on in Chapter 1: the growth of connexity, an increasing reversibility of forms of mobility and the advent of the ubiquity of actors. The terms used to qualify the three processes are seldom used in the social sciences due to the fact that they refer to phenomena that are often confused; so much so that differentiating between them also calls for clarification.

First of all I will define the pairs 'contiguity-connexity', 'irreversibility-reversibility', and 'unity-ubiquity'; then explain the phenomena to which they refer, with the aid of relevant literature, and finally discuss their impacts and the debate surrounding them. From this, I will suggest some counterpoints to the idea of social fluidification.

Before embarking on the real analysis, however, it is important to stress that, whilst certain aspects of mobilities are very well documented (like their spatial implications and migrations in general), there are serious gaps with respect to other aspects. This is probably due to the disciplinary partition of research areas, particularly with regard to the apprehension of time and the underdevelopment of transport sociology. Concerning time, much research limits the study of mobility to its spatial dimension, neglecting its temporal dimension (Ascher 1998: 141-161). The latter refers to constraints on activity programmes of everyday life and also to social rhythms and to the different dimensions of the life course (family, occupational, leisure). The underdevelopment of transport sociology constitutes a large handicap for the understanding of the process of spatial mobility (De Boer 1986). This lack could be explained by the fact that until recently mobility has been considered as a neutral interstice between existing activities (Urry 2000a: 39). More specifically, the neglect of the automobile by sociology (Urry 2000a: 58; Sheller and Urry 2000) has created a void in sociological analysis and prevents certain types of analysis on urban research. The review of the literature on which this chapter is based, therefore, also mirrors these limitations.

Three phenomena from which the compression of time-space originates

From contiguity to connexity

The notions of contiguity and connexity come from the vocabulary of human geography and refer to the way in which the co-presence of actors is established. Connexity can be defined as the establishing of relations using the intermediary of technical systems. In the case of contiguity, on the other hand, this relationship is established by spatial proximity.

Connexity allows the interaction of actors by cancelling out spatial distance. Advanced communication systems as well as fast transport, like aircraft or the TGV on an inter-urban level and the RER (regional high-speed express network) or the automobile on an intra-urban level, are the vectors of connexity, which is characterised by a 'tunnel effect'. The appropriation of space crossed between the origin and the destination is not possible. Countryside passes by the window of the TGV at 300 kilometres per hour, or on the other side of the windscreen of a large saloon car on the motorway; towns appear and disappear in an aeroplane window without it ever being possible for us – or even our eyes – to linger as everything passes by so quickly.

In contrast, contiguity assumes a continuity of the possibility of appropriation of routes taken. Contiguity relates to the traditional way people relate to one another in a city, town or village and implies density.

New means of telecommunications and rapid transport have allowed connexity to develop, with actors now fitting into several spatial patterns at the same time (Offner 2000: 172). The existence of the development of connexity is widely accepted. Its interpretation, on the other hand, is the subject of controversy. I will highlight two examples:

The end of territories. Certain researchers are of the opinion that connexity has taken on such dimensions that it is putting an end to contiguity and territories (Badie 1995, Couclelis 1996, Balandier 2001). For these authors, social integration comes from connexity. Thus, there are no longer any territories, these having been based on a principle of homogeneity that connexity has dispersed. By extension of this reasoning, there are some who have even predicted the end of towns and cities: do we still need the spatial density and proximity that define a city if all distance can be crossed so quickly? Melvin Webber or Francoise Choay, for example, believe that the town and the city as morphological realities have now been replaced by a lifestyle (Choay 1994). For other researchers, the end of contiguity and cities has not been established. Several of them have pointed out the irony that the imminent death of the city has been predicted for the last thirty years and we have yet to see any signs of it. In support of their scepticism, they highlight the attachment of inhabitants to their home territory and point to the voluntary persistence of numerous practices of daily life carried out in spatial proximity. In short, they show how contiguity and connexity have a tendency to merge (Bellanger and Marzloff 1996, Remy and Voyé 1992). People covering long distances during the course of their daily lives often conduct activities near to or in their own neighbourhoods. The most significant example of this is undoubtedly the back-to-back nature of daily activities. Transport surveys show that car (and sometimes public transport) journeys are increasingly used as opportunities to combine micro-activities (Bellanger and Marzloff 1996). The establishment of new settings for closeness, like the automobile, which become places of sociability – enclosed spaces somewhere between public and private space where one can chat (Bordreuil 1997, Bellanger and Marzloff 1996), also highlights this tendency.

The growth of actors' freedom. Many authors consider the phenomenon of connexity in relation to social integration as a factor contributing to the broadening of the realms of the possible in conducting one's life. Then again, empirical research can neither confirm nor deny such affirmations. This is the second subject of controversy that I would like to tackle. One point of view well documented in literature (Ascher 1995, Chalas 1997, Lévy 2000, Remy 2000) is that the increase of connex mobility could be considered as a broadening of the possibilities of individual choice. In the conclusion of 'La ville emergente', Yves Chalas shows that this process leads to what he calls 'the city of choice':

> The principle of free choice (...) manifests itself in the fact that (...) more and more inhabitants not only construct their social networks but even make their unusual/special purchases and those relating to their daily nutritional needs, use services, whatever these be, of an institution, doctor or dentist, or spend their free

time where they see fit, very far or very close to their homes, it is of little importance, and not necessarily or principally in their municipality, by virtue of an attachment or loyalty to or an identification with this municipality (Chalas 1997: 263) (my translation).

On the other hand, the growth of connexity could be viewed as the emergence of a new form of spatial constraint, having the tendency to render that which is spatially close remote, especially by urban divisions (Heran, 2000).

Other authors go even further, developing the idea that the growth of connexity is such that it reduces 'social capital' (in the sense of social relations) by means of the protraction of travel time that connexity implies. Robert Putnam is one of these authors:

> The car and the commute are demonstrably bad for community life. In round numbers the evidence suggests that each additional ten minutes in daily commuting time cuts involvement in community affairs by 10 percent – fewer public meetings attended, fewer church services attended, less volunteering, and so on. In fact, although commuting time is not quite as powerful an influence on civic involvement as education, it is more important than almost any other demographic factor. And time diary studies suggest that there is a similarly strong negative effect of commuting time on informal social interaction (Putnam 2000: 213).

Nevertheless, on this last aspect, some researchers cast doubt on the links between social capital and connexity. They show that in certain European countries the growth of the automobile is congruent with a strong revival of memberships of clubs, etc. and of creations of local societies (Forsé, 1999).

From irreversibility to reversibility

The notions of reversibility and irreversibility are borrowed from astrophysics and have only recently entered the language of the humanities (Rapoport 1996: 271). According to Schuler et al (1997), reversibility and irreversibility are defined with reference to the impact of mobility on actors' identities. Irreversibility is a total social experience. It is an experience to which one is obliged to be faithful. A good example of one of the most irreversible forms of mobility is certain forms of international migration like the boat people, which implies a change of cultural context, an experience one cannot escape from in the sense that such people are immediately identifiable in terms of racial or ethnic differences. Another example of an irreversible mobility scenario is that of enforced residential mobilities such as going to prison. By the stigma attached to it a prison sentence marks the identity of an individual definitively and follows an individual all his or her life. It is important to beware of deducing from these two examples that irreversibility is only the expression of a constraint. One can perfectly well imagine where irreversibility is chosen, such as in the case of someone deciding to set up his or her life elsewhere.

Reversibility, on the other hand, is a social experience that can be cancelled. The most reversible mobilities are those that one does not fully remember. Often repetitive, they refer to the realm of the non-event. Commuting and business trips are good examples of this: a person goes to work every day, but cannot recall exactly each daily journey; an international consultant travels a lot but does not recall each individual flight. One cannot, however, assume from this lack of precise memories that the most reversible mobilities are without impact on identity. Indeed, their very repetitiveness often makes them builders of identity in both the individual's own eyes and in the eyes of others. Conversely, the day these experiences cease they leave no trace on the identity, in contrast with irreversible mobilities.

Nevertheless, reversibility and irreversibility must be considered as ideal-types in that forms of mobility are never purely reversible or irreversible. The reversibilisation of spatial mobility has also caught the attention of researchers (Cwerner 1999, Urry 1990, Wiel and Rollier 1993), although less has been written about it than the growth of connexity. The following three phenomena stand out in relation to this:

1. I have noted the phenomena of substitution of the most irreversible forms of mobility (migration, residential mobility) for more reversible forms (daily mobility, travel). Typical of this is the use of the speed potential provided by motorways to live farther away from the work place, thus avoiding relocation (Wiel 1999, Putnam 2000). This substitution implies a transformation of mobility temporalities from long to short term. Above all, this substitution corresponds to a modification of the impact of mobility on identity. By travelling rather than migrating or by commuting rather than moving, because this implies frequent return, one preserves one's identity of origin and all its social networks. In effect, this protects the actor from all the effort of repositioning and reconstructing an identity which a migration or a move imposes. Colin Pooley and Jean Turnbull's recent research (1998) on the history of mobility in Great Britain illustrates reversibilisation by substitution, with the additional backing of historical analysis. The data collected shows, in particular, a substitution of migrations for residential mobility perceptible since 1880 and accentuated from 1920 onwards (Pooley and Turnball 1998: 72). It appears that this form of substitution is effectively a process that started with motorised transport and developed alongside its expansion.

2. The 'reversibilisation' of the different forms of mobility themselves is significant. More than in the past it is possible to cancel the impacts of spatial distance. This aspect is related to connexity. For example, international

migrants are able to keep in touch with their families and friends through communication networks (Cwerner 1999, Stalker 2000: 120, Stimson and Minnery 1998), as rapid forms of transport allow both easy visiting of migrants and for them to travel. Residential migrants continue with their old habits in their previous neighbourhood, as Marc Wiel and Yann Rollier's brilliant research on peregrination in the urban district of Brest has shown (Wiel and Rollier 1993). Another example may be the tourist gaze, which universalises places and reduces confrontation with the unknown (Urry 1990). In this way the journey is reversibilised as visited sites become more and more like standardised products that are known to us before we see them.

3. Thirdly, I note that most reversible forms of mobility are increasing in terms of distance and becoming more reversible. This tendency has been constant since the 1950s for daily mobility (Salomon et al 1993) as well as for tourism, with increases in both international tourism (Urry 1990) and the development of short-term urban tourism (Potier 1996). Thus, between 1970 and 1993 in the countries of the European Community, there has been an increase from 2000 to 4000 billion kilometres travelled per year, most of which has been covered by automobile (European Community, 1995). This equates to a constant acceleration in the speed of journey times. I can add to this that the choice of transportation is emphasised by this process. The automobile is more and more widely used for daily journeys. The use of cars makes drivers and passengers invisible to others (Whitelegg 1997) and thus more reversible than if they used other means of transportation.

As with the growth of connexity, although there may be consensus about the phenomenon itself, the impacts of this growth of reversibility are the subject of diverse interpretations (not always contradictory, however):

• Certain analysts see the end of mutual engagement in the growth of reversibility. For these authors, the increase in reversibility confirms the idea that not only is spatial structure supported by flux but that mobility also allows escape. The increase in reversibility permits a form of disengagement vis-à-vis local societies: one is there without really being there, ready to seize any opportunity for mobility if it increases one's well-being. The examples that illustrate this idea generally come from the economic sphere. Thus, there are works that show how certain businesses have very weak links with their location, and that if the conditions seemed better elsewhere they would move (see for example Friedman 1995). Other works show that the attachment of employees and management to their company is less than in the past, something that could, moreover, be considered as a consequence of the previous point.

- Other researchers set against this interpretation the fact that the reversibilisation of mobility produces new fixities that actors are more and more reluctant to disengage themselves from. From this second viewpoint, reversibilisation is interpreted as the effective disappearance of the unknown. If in the past spatial mobility necessarily meant a leap into the unknown, with reversibilisation this is no longer the case; one can travel thousands of miles while remaining in a completely familiar realm (Urry 1990). Confrontation with the unknown becomes all the more worrying, and is avoided because of the perceived risk attached to it.

- Numerous authors consider that the reversibilisation of some forms of mobility is the sign of the advent of a reflexive society. In this third view, the reversibilisation of mobility is an integral part of a larger phenomenon by which actors reinterpret and readjust their social practices in the light of information related to these practices (Beck Giddens and Lash 1994, Asher 2000). The capacity to analyse the consequences of what is happening *while* it is happening is characteristic of reversibilisation, thus maintaining the possibilities of withdrawal from a decision at the time of each mobility. Researchers have studied the automobile from this point of view. They interpret its success in relation to forms of public transport in terms of the flexibility offered by the car in reorganising mobility and activities during the course of the day.

From unity to ubiquity

These two notions derive from general sociology. Unity refers to the idea of the localized individual actor forming a coherent whole with an identity and a culture. Ubiquity refers, on the contrary, to a plural actor characterised by the multiplicity of his or her roles and identities and the possibility of acting at a distance.

The development of ubiquity refers to the relationship between spatial mobility and social mobility. In a model of society characterised by the separation of functions in social space (sexual division of work, predominance of socio-professional status in identity) and spatial space (functional specialisation of land), spatial mobility goes hand in hand with social mobility. A change of roles generally implies a change of location. The boundaries of this model, however, have become blurred to give way to an increase in roles and their spatial superposition. Thus, spatial and social mobilities interact differently. The progressive wearing away of the sexual division of roles ('working women', 'new' fathers etc.) and the development of free time multiply horizontal social mobilities without this necessarily being accompanied by spatial mobility. Thus, for example, the home is now more and more likely to be a space for leisure (video, TV, Internet) or a place of work (thanks mainly to computers linked to the Internet) as well as being a domestic, family space. The result of this is a jumbling of public and private spheres and a merging of free time and worktime. Ubiquity is largely linked with the compression of time-space procured by telecommunication technologies.

By compiling syntheses of research studies (Lahire 1998, De Singly 2000, Tarrius 2000) it is possible to distinguish several forms of ubiquitous mobilities associating time, space and identity.

The superposition of daily life. People often use the speed provided by technical systems of transportation and communication to increase the speed of succession in spheres of activities of daily life. The consequence of this fact is the fragmentation of these spheres of activity or even their temporal superposition (Hochshild 1997, Jurczyk 1998). Many examples can be cited in support of this first form of ubiquity, starting with the development of work carried out during journeys, an activity made possible by mobile phones and laptop computers. The fragmentation of spheres of activity also refers to the interruption of one activity by another, a situation that is more and more common in the era of mobile phones (Ascher 2000: 206). But one of the most significant examples of ubiquitous superposition is, without a doubt, the development of the surveillance video in crèches in the United States, which allow parents to see their child at all times.

The intensification of social multi-belonging. Ubiquity also takes the form of multiple social identities. This is undoubtedly the most debated form in sociology and is very frequently expressed in interviews: 'I am white, Jewish, from a well-to-do neighbourhood, son of a lawyer, I'm a man, a husband, a father, a teacher, a colleague – I am all these things at the same time' (Ostrow 1990: 81), 'I am speaking to you as a woman ... a young person, as an occitant' (from a region in southern France) (Dubet 2000). Multi-belonging is linked to the superposition of spheres of daily life (Bailly and Heurgon 2001: 13-51). The copresence of several roles to play in the same space implies being able to move very quickly from one role to the other.

The development of identity with more than one spatial reference: being from different places at the same time. The increase in the number of migrant people constructs a new culture of mobility and produces multicultural belonging. This leads to a development of 'New Diasporas', in which identities are constructed from migratory experience. These people are able to develop multiple identities according to their mobility experience. They often feel themselves to be from different places (Rapport and Dawson 1998). As Alain Tarrius states, on the basis of abundant research on the evolution of migratory forms: 'Diasporas and nomadic groups, the time of necessary occupation of territories by patience or the violence of an invasion of local societies in a "make way for us!" style has passed. To be from here and from over there at the same time is not a rhetorical image: it is a brand new status established by knowledge, slipping to, keeping close to, being near to others, all the others, and avoiding subjugation as much as possible' (my translation) (Tarrius 2000: 243-244).

As with the other two phenomena the meaning of the concept of ubiquity enjoys a large consensus while its social implications are much debated. Ubiquity is sometimes considered as the sign of a virtualisation of the world, placing the desires and aspirations of individuals faced with a quasi-unlimited domain of possibilities at the centre of the constitution of the social. This type of analysis is based on the fact that the different forms of ubiquity have only been able to develop thanks to modern information and communications technology. In this view, mobility is no longer spatial (being here or there amounts to the same), but social and virtual: there are those with access and those without access. It is no longer geographical space that differentiates but virtual space (Knox and Taylor 1995). Ubiquity allows the smoothing out of a number of the inequalities between central and peripheral regions or between countries of the north and of the south. It is even presented sometimes as a substitute for land-use planning, although this thesis is hotly disputed.

For many analysts, this notion of a virtualisation of the world is akin to a form of technological fanaticism. The idea of a virtualisation of the world by networks implies the replacement, without loss of quality, of co-presence by on-line relations (Boden and Molotch 1994, Offner 2000). The main argument against this interpretation is that technological telecommunications systems are above all used for short-range communication. Thus the Internet seems more to be a medium for the reconquest of local sociability than that of contact with an unknown person in Singapore (Rallet 2000). Moreover, several researchers have shown that the more telecommunication there is, the more social mobility. Thus, telecommunication generates a need for co presence. When one gets to know someone through the Internet the need to meet is felt from the moment the relationship takes off. In other respects it seems that the ubiquity procured by telecommunications brings with it stronger and stronger locational constraints more for companies and that these probably contribute to the increase in socio-spatial inequalities (Offner 2000). This last aspect, moreover, derives largely from two preceding ones: if new information and telecommunications technology is mainly used locally and they imply seeing one another face to face, then these technologies do no more than accentuate spatial inequalities. The social science literature concerning metropolisation is categorical on this point. The phenomenon of the concentration of powers of economic determination that the biggest agglomerations enjoy can be explained to a large degree by their good transport links and by the need for the spatial proximity that network activities imply (Veltz 1996).

Counter-arguments to the theses of social fluidification

The growth of the connexity, reversibility, mobility and ubiquity of actors are established facts. The impact of these phenomena on the social, however, is the subject of diverging analyses. In a general way the examples developed in the preceding sections suggest that these divergences are largely linked to the consequences of the transformation into mobility of the speed potentials offered by systems of transport and telecommunication. There are generally two opposing interpretations: one that is based on the principle that potentials for speed are used (or will be) to 'escape', and the other which considers that the appropriation of these potentials is not always effective and is subject to multiple constraints. Thus, for certain analysts the possibility of connexity spells the death of territories and increases the degree of liberty of the actor, whilst others feel that this possibility encourages spatio-temporal re-compositions which do not necessarily go in the direction of fluidification. Similarly, the possibility of reversibilisation is interpreted by some as the end of mutual engagement, whereas others see in it the advent of a reflexivity made up of contingencies. Ubiquity, then, can equally be interpreted as the dawning of an egalitarian virtual world or, on the contrary, as a possibility only making sense in relation to pre-existing socio-spatial structures. It is on these two interpretive levels that the current debate seems to stumble. For lack of an adequate concept allowing the analysis of the transformation of speed potentials into actual mobility, researchers too often take refuge behind implicit presuppositions that oscillate between reification and technical determinism. Four questions in particular can be identified:

- Are speed potentials used to frequent a territory or escape from it?
- Are speed potentials systematically appropriated or not?
- Is access to speed potentials subject to multiple constraints or is it, on the contrary, very free?
- Do speed potentials tend to diminish socio-spatial inequalities or augment them?

These questions, which largely remain without definitive responses, suggest that speed potentials may just as well produce fluidity as fixity. For this reason, they allow the formulation of several counter-arguments to the theses of fluidification of the social presented in Chapter 1 based on research showing that the appropriation of speed potentials is not always effective and is subject to constraints.

Networks and territories make up a system

The first counterpoint consists in affirming that connexity, reversibility and ubiquity can only make sense in relation to the territories and social structures to which these phenomena are subject. It is therefore false to consider rapid transport and telecommunications as capable of creating a fluidification of the social. If fluidification exists it comes from somewhere else. Jean-Marc Offner (2000) demonstrates well the need to keep the novelty of transport and telecommunication networks in perspective and to beware of analysts who consider that territories can to a very large degree be determined or even cancelled out by these networks. After all, there were discourses on the impact of the telegraph and roads on territories a hundred years ago predicting their imminent dissolution. From this point of view, territory determines these networks a priori as much as networks fluidify territories. There are three specific points that emerge from the literature:

- The first is that network services have differential performances. There is not, therefore, a homogenised flux space but rather diversified spatial dynamics. It is therefore futile to seek to explain interactions between networks and territories without taking into account modalities of conception and management of these networks. As such, it is not possible to consider networks as unilateral instruments of possible deterritorialisation. Networks adopt the existing spatial structures and especially relations between centre and outskirts. Moreover, they are often only available at certain points and not continuously.
- The second refers to the structural effects of networks on territories. Networks do not have structural effects on territories in terms of the localisation of activity. The structural effect thesis comes up against three types of criticism: methodological, empirical and conceptual (Offner 2000: 169-170). Methodologically, before-after evaluations do not correctly measure the spatial impact of a network, confusing, as they do, co-occurrence, correlation and causality. Empirically, evaluations show that speed potentials have the effect of accelerating pre-existing tendencies. From a conceptual point of view it is not possible to base a causal analysis on just one factor such as technical innovations.
- The third refers to appropriation. The usage of networks is sometimes defined by territories, which then have an effect on flux (Rallet 2000). A significant example of this is undoubtedly that of workers on the border between Switzerland and Germany and between Switzerland and France. If they cross the national borders this is not because of transport networks (which remain of a fairly mediocre quality), but because the salaries are higher in Switzerland and Switzerland needs skilled workers.

Connexity, reversibility and ubiquity as constraints

The third counter-argument consists in highlighting the fact that connexity, reversibility and ubiquity are as much the result of a system of constraints as of a broadening of possibilities of life choices. The appropriation of speed potentials is, therefore, strongly conditioned. The force of these constraints suggests that the contemporary mobile world is no more fluid than the one it has replaced. This counterpoint could be illustrated in many ways, notably by referring to the research on the erosion of spatial proximity or, in another area, the flexibility imposed in the workplace (temping and short-term contracts, etc.). But research into the way people conduct their lives seems to me to be more convincing using the support of this thesis.

In Germany, a branch of empirical research has developed since the 1980s around the notion of 'Alltägliche Lebensführung', which could be translated as 'daily life management' (Kudeva and Voss 1995). In this view, everyday life is defined as all activities that make up daily 'routine'. This branch of research aims to investigate the degree of flexibility existing between different activities thus defined. It is thus a matter of identifying and describing the way these activities are organised. According to Voss, the way one leads ones life depends on the opportunities that emerge from ones social context. It thus depends on the objective constraints to which individuals are subject in their daily lives and on multiple influences of a sociocultural nature. Nevertheless, the basic idea of research concerning daily routine is that individual actors have a degree of freedom, as even with the presence of extremely rigid social contingencies, actors must function as individuals.

In concrete terms, research about the way people conduct their daily lives suggests that connexity, reversibility and ubiquity are often the result of a system of social and spatial constraints which impose themselves on the individual and not as a broadening of their degree of liberty. The example of working women developed by Karin Jurczyk (1998) is a good illustration of a system (non areolar), which constrains connexity, reversibility and ubiquity.

'Working women devise a "timetable" for the entire family, as well as for its individual members. They remember everyone's appointments and try to coordinate that variety into a harmonized unity to enable the family to spend time together. These timetables often look quite complicated. They take account of school and working hours, children's regular afternoon appointments such as sports or music classes, doctor's appointments, visits to events and friends. Also included in these timetables is the time it takes to travel between the different locations, a factor that is increasingly unpredictable on account of changes in the organization of public and private space' (Jurczyk 1998: 295).

Fluidification affects at best a minority

The last counter-argument consists in recalling the very localised nature of the phenomena of connexity, reversibility, and ubiquity of forms of mobility, which concern but a few social categories in the western world. Consequently it consists of affirming that the evolution of forms of mobility corresponds to an accentuation of social and spatial inequalities with the result that it is not possible to talk about a fluidification of the social. The reinforcement of big centres to the detriment of rural areas or medium-sized towns, the rift between countries of the North and countries of the South, the persistence of inequalities of gender, race and the growth of social inequalities are all as much signs of this state of affairs.

To talk about social fluidification in this context thus refers at best to a small section of the population: men, white, and young, western, urban, with a high level of education. The other possibility is that it is a manifestation of a form of ethnocentrism on the part of researchers who generalise their own situation.

This type of argumentation is often put forward in Anglo-Saxon gender studies and in research into racial discrimination. In the area of migration it has also been put forward to demonstrate the occidental character of approaches used to deal with migration questions in emerging countries (De Haan 1998). This 'theoretical imperialism' is sometimes denounced as a barrier to the advancement of social science literature on mobility in emerging countries, imposing a frame of reference inappropriate to these countries (De Haan 1998: 3-4).

Beyond the limits

Though the growth of connexity and the reversibilisation of forms of mobility and ubiquity are now regarded as established facts that are hardly discussed any more, the lines of debate concerning their societal impacts are, on the other hand, numerous. This controversy results in three important conclusions.

The first is that debate relating to the impact of connexity, reversibilisation, mobilities and ubiquity gets bogged down as a result of confusion between speed potential and effective mobility. The normative dimension of this confusion is, moreover, constantly aired in the form of technological fanaticism postulating that the full potential of speed offered by transport and communications networks is actually used and allows escape from structures and territories. I deduce from this the confirmation of the importance of having a conceptual apparatus capable of re-defining and being more precise about the notion of mobility.

The second is that mobility and fluidity do not necessarily go hand in hand. Numerous research papers suggest that connexity, reversibility and ubiquity by their nature constrain actors' actions and thus do not correspond to a broadening of their degree of freedom.

The third is that it is possible to formulate counter-arguments to the theses of fluidification of the social that do not necessarily make reference to the areolar model. Speed and mobility potentials are carriers of constraints of access and appropriation that do not necessarily originate from a categorical belonging (as research on the way people conduct their lives shows). Taking the debate further and creating a dialogue between research findings and the areolar, network, liquid and rhizomatic models is the conceptual task which will make up the heart of this work. This is the focus of Chapter 3.

3 Re-thinking Spatial Mobility

This chapter will aim to develop a conceptual tool to measure mobility, and for this purpose it has been divided into three sections. The first discusses the limits of the concept of mobility. The second is devoted to the development of a conceptual device to be used to surpass these limits. The third section proposes to evaluate the significance of spatial mobility as a possible vector of social fluidity based on the proposed conceptual device.

Limits to be surpassed

In Chapter 1, I expressed regret at the lack of conceptual tools with which spatial mobility could be measured. This is notably the result of two complementary aspects: the polysemic notion of mobility, and the insistence that spatial mobility must focus on the geography of movements. Conceptualising mobility implies going beyond these two limits.

A polysemic notion and piecemeal knowledge

Spatial mobility is not a univocal concept, but rather has different meanings. It can refer in one context to physical movements, and in another to communicator metaphorical movements, and concerns humans, goods, information, ideas and so on (Urry 2000a, Bauman 2000, Castells 1996). Spatial mobility is thus an exceedingly general and vague concept.

At least four meanings are currently used in the social sciences (Schuler et al 1997) to describe the mobility of people alone: residential mobility (with reference to residential cycle), migration (international and interregional immigration and emigration), travel (tourism and business travel), and daily mobility (daily journeys such as commuting).

This profusion may in certain cases be an advantage, because it avoids a single connotation and permits an appropriation of different kinds of theoretical perspective. On the other hand, it is also – and above all – a serious limitation on progress in knowledge about mobility, for several reasons.

How can we describe phenomena with precision using an imprecise notion? When a geographer refers to mobility, he/she is not speaking of the same thing as an engineer or sociologist who borrows this concept, which makes dialogue between their respective branches difficult. The bottom line is that when we refer to mobility, we do not know precisely what we mean; it all depends on the branch of study we are in. In today's world, progress in any field necessarily implies consulting different branches of literature. Not to do so would be to miss out on a considerable potential source of enrichment.

The different accepted definitions of mobility shape research. Each definition that relates to one aspect of spatial mobility refers to a specific field of research that deals with a specific object. This situation means that it is nearly impossible to deal with transversal subjects of research without running into these different definitions. For example, we know very little about the phenomenon of dual residence, which straddles the areas of residential mobility, travel, and daily mobility. On the other hand, each of these forms of mobility has caused much ink to flow, yet at times it seems that no progress has been made; we repeatedly undertake research forays into the same old sectoral topics such as the setting of habits related to daily modal practice in lifestyles, residential mobility and the life cycle, modal practice, without there being any discernible scientific progress. Yet, it is the interconnection of these different forms of mobility and the debates conducted by the actors that fully reveal the phenomena of mobility and their implications. The analysis of the three stages of growth of connexity among the forms of mobility, of their reversibilisation and their ubiquity shows the importance of this interconnectivity, because it points specifically to the combination of the different forms of mobility in terms of reinforcement, substitution, and rhythms.

In sum, spatial mobility is not a concept that belongs to social sciences, but a piecemeal notion scattered among different fields of research and disciplines. It is also vague because it covers different phenomena with no links between them. This leads to a neglect of transversal subjects of research.

A focus on the geography of movements

The notion of mobility focuses the researcher's attention on movement in space-time rather than on the actor. This may be another element of explanation for the non-interest of many sociologists in this notion: they link mobility to human geography because of this focus on the movement itself. Yet, actors are central in the mobility process. It is well and truly the aggregation of actions that permits the development of technical systems of transportation and telecommunication, and then their appropriation by the use of the speed potential they provide. In addition, with the multiplication of opportunities available, movement in geographical space is no longer a contraint for communication. Basing the analysis on spatial mobility, therefore, prevents us from studying the relationship between mobility and communication.

This insistent focus is at the origin of the confusion between mobility and fluidity which I regretted in Chapter 1, and it opens the door to hasty conclusions and suppositions. If one approaches the geography of flows in isolation without looking at the logic propelling them, this leads to the development of regulatory analyses that take as their point of departure that increased speed of travel and greater distances covered are synonymous with freedom. The intent behind the movement is too often depicted as a simple equation summed up thus: mobility is good, because it equals open-mindedness, discovery, and experience, and an effort must be made for individuals to maximize mobility for this reason. This affirmation is part of a value system and illustrates that spatial mobility is not devoid of connotations from a social point of view.

The speed potential permitted by the technology of transport systems is often seen as an instrument for offering people mobility, as a means to make them mobile. It is imperative that this confusion between the speed potential offered by transport technology and the unquestioned attribution of strategies to individuals and to people's mobility be abandoned if we are to study spatial mobility from a sociological point of view. Only by integrating the intentions of people and the reasons which make them mobile or which, on the contrary, leave them immobile will we succeed in attaining this goal.

Getting past this confusion suggests redirecting the interest of researchers towards the aspirations and plans of those involved, as well as the things that motivate them, and their possible realm of action.

Conceptual proposals

In order to overcome these two barriers outlined in part 1, I propose to invert the research perspective and to take instead as the point of departure people's potential for mobility by expanding on the two concepts of motility and mobility.

Motility

The introduction of a new term into scientific discussion can be justified from different standpoints. A new term may be used to identify a previously unknown phenomenon that does not correspond to any existing definition. But a new term can also appear as necessary to describe an existing phenomenon from a new angle, and in this case one speaks of reconceptualisation. Finally, a new notion may be required to fine-tune a concept. Like many new words that appear in the social sciences, motility is the product of reconceptualisation and of fine-tuning.

Motility can be defined as the capacity of a person to be mobile[1], or more precisely, as the *way in which an individual appropriates what is possible in the domain of mobility and puts this potential to use for his or her activities*. Several authors have recently argued the importance of this perspective for mobility research. This is the case of Jacques Lévy, for example, who defines three virtual components of mobility: possibility, competence and capital (Lévy 2000), all three of which are part of motility and constitute the beginnings of conceptualisation. Jean Remy also evokes mobility as an appropriable resource (Remy 2000). The common key idea of these considerations is that every actor has her/his own potential for mobility, which can be transformed into movement according to aspirations and circumstances.

Adopting the notion of motility does not imply a reification of space-time: it only makes sense if there is something to appropriate, which means a context and access to this context. In this perspective, I concur with Andrew Sayer when he writes that 'space makes a difference in terms of settings or contexts, for while institutions have spatial extension and perhaps a particular shape and degree of mobility, they also are set in a spatial relation to other objects: social processes do not occur *tabula rasa* but "take place" within an inherited space constituted by different processes and objects, each of which have their own spatial extension, physical exclusivity and configuration' (Sayer 2000: 115). Similarly, developing an approach with respect to motility does not mean denying the structural or cultural dimensions of an action. People's capacity to be mobile is linked to these dimensions; it remains to be seen how, and whether or not this has changed with respect to the past.

Taking inspiration from Lévy's work (2000), I will consider that motility is comprised of all the factors that define a person's capacity to be mobile, whether this is physical aptitude, aspirations to settle down or be mobile, existing technological transport and telecommunications systems and their accessibility, space-time constraints (location of the workplace), acquired knowledge such as a driver's licence, etc. Motility is thus constituted of elements relating to access (i.e. available choice in a broad sense), to skills (the competence required to make use of this access) and appropriation (evaluation of the available access).

Access. Access refers to the range of possible choices in a place, and is comprised of networks and of flows, of territories and of places. It has two components: options and conditions. The options are comprised of the whole range of means of transportation and communication available and the whole range of services and equipment potentially accessible in a given time unit. The conditions refer to the accessibility of the options in terms of price and schedule. Moreover, the components of access depend widely on the spatial distribution of the population (for example, small towns and big cities provide different 'ranges of choices' in terms of available equipment and services), on the sedimentation of spatial policies especially for localisation and transportation accessibility, and on social inequality (especially in terms of purchasing power).

Skills. Skills refer to the *savoir-faire* of those involved. Three aspects are central to the skills component of motility. The first concerns the physical abilities that mobility implies, i.e. the ability to walk, to see, etc. The second concerns acquired skills that allow one to be mobile, for example a driving licence or the knowledge of the English language for international travel. The third comprises organisational skills, such as the way in which activities are planned, and involves researching information, spontaneity, and so on. Skills are thus multi-faceted and have to do with a person's age and point in his or her life course.

Appropriation. The aspect of appropriation is how people interpret access and skills. Appropriation is shaped by the aspirations and plans of individuals and thus stems from their strategies, values, perceptions and habits. It is this aspect that people will use to judge whether or not access is appropriate and thus to be taken into consideration. Appropriation is also the means by which skills are evaluated and decisions made as to whether or not they are worth acquiring. Appropriation is constructed through the interiorisation of standards and values, and as such has to do with gender and the point reached in a person's life course.

These three aspects function in tandem and, as the following two examples illustrate, are inseparable:

- *Individuals evaluate technological transportation systems in such a way as to influence public and private investment and thereby influence access and skills. A good example of this is the renewed development of tramway networks in France.* The modern and urbane image (appropriation) of these trams allowed public transport to cease being perceived as a means reserved only for those who are 'captives' of the city (skills). The positive image associated with the tram (appropriation) also led to an inflow of public investment in large French cities (access);
- *The introduction of new potential for mobility in a given context* (for example a new, low-cost airline company or the creation of a car sharing cooperative) *can make room for the expression of an existing but latent demand and helps to modify access and skills. If users judge the new service favourably (appropriation), they will be likely to use it, by means of the required financial resources (access) or the acquisition of the necessary skills (such as the purchase of a computer so that tickets can be reserved via the Internet).*

The three dimensions of access, skills, and appropriation together constitute a propensity to be mobile which is motility, and which is likely to vary in intensity from one person to another. Furthermore, this propensity for mobility lends itself to some forms of mobility more than others. For example, it is possible to have a very high propensity for travel while having very strong residential roots.

Although motility is identified on an individual level, it is not formed individually – in fact, quite the contrary. The foregoing examples show this by illustrating how motility is eminently social. But these links run even deeper. Motility is in fact formatted by the life course of those involved and by their financial, social and cultural capital, which together define the range of possible specific choices in terms of opportunities and projects. With respect to the impact of one's life situation on motility, it should be specified that the family situation and composition of the household play a central role, because they imply complex compromises of the individuals involved: the differing degrees of motility of the members of a nuclear family with respect to access often require creative solutions for planning daily mobility, especially with respect to children.

Mobility

Approaching movement via motility does not, however, dispense with the need to reflect on the notion of mobility, even if only to examine the reasons why understanding has been so fragmented.

Of the numerous researchers who have argued in favour of developing a single global concept of spatial mobility (Brulhardt and Bassand 1981, Schuler et al 1997), none has yet produced a truly complete formula. Although it now appears to be generally accepted that mobility can be considered as a total social phenomenon in the sense defined by Marcel Mauss and that the different forms of spatial mobility form a system, constructing a model of the interaction between the different forms of movement is deemed so complex by some authors that they wonder whether such model-building would make a contribution to research in operational terms (Tarrius 1989). And yet the value of a concept is often its simplicity, and this is even truer for a general notion like mobility that concerns many different topics. Another constraint of the systemic approach to mobility is that it is too mechanical. According to Jean Remy, 'The quality of a human system is to be imperfect from the point of view of systemic logic' (Remy 2000: 176); in other words, not only are there interdependent relationships, but also interactions that escape the system. Finally, the main virtue of a systemic approach to mobility is that it considers spatial movement *tout court* as a phenomenon that can take alternative forms. Such an approach rebuilds unity where research fields and disciplines scattered the pieces of the puzzle.

I therefore propose to consider spatial mobility as a phenomenon that revolves around four main forms: migration, residential mobility, travel, and daily mobility. These four forms are interconnected and each linked to specific social temporalities: the day and the week for daily mobility, the month and year for travel, the year and life course for residential mobility and life history for migration. These different forms all have an impact on each other. The forms of mobility that are linked to the longer temporal durations (life course, life history) have a systematic impact on the shorter forms. After moving, people inevitably have a daily mobility that is different, even if only because the pace followed on a daily basis has changed; an international migration not only modifies a person's daily mobility, but may also generate travel (to see old friends or family members left behind) and may also create specific residential mobility (an initial move into a furnished apartment is followed by the purchase of another), and so on.

The four forms of mobility can be classified according to two aspects (see Table 3.1), which are the temporal categories to which they are linked (long or short temporal duration) and the space in which they occur (internal or external to the area in which people live)[2]. The combination of these different forms of mobility constitutes the space-time expression of lifestyles and can be measured in terms of rhythm.

Table 3.1   The four main forms of spatial mobility

	Short duration	Long duration
Internal to the living area	Daily mobility	Residential mobility
Near the outside of the living area	Travel	Migration

The space-time expression of lifestyles as developed in Chapter 2 is currently being modified through three growth processes of connexity, reversibilisation of the forms of mobility, and ubiquity. This is bringing about the emergence or the development of new forms of mobility, which fit in between the four main forms, as the following examples show:

- Dual residence (fits in between daily mobility and interregional migration or residential mobility). Although the concept of two seasonal homes[3] has long existed, its development on a weekly scale is new and, what is more, spans very different situations. In couples where both partners work, the fact that both jobs are not located in the same city often results in this kind of compromise (Lévy 2000). Dual residence also exists where second homes are inhabited three days per week (Viard 1995). Joint custody of children following a divorce is another, growing example of dual residences (De Singly 2000: 219-235);
- When the workplace is very far from home, a practice other than dual residence is currently in development: very long-distance commuting that is combined with working from home (this fits in between daily mobility, residential mobility, and travel). In this form of mobility, the person travels to the workplace only one or two days per week and works at home

the rest of the time (Hochshild 1997). This practice relies heavily on the capacity for long distance work permitted by electronic mail;
- Separate homes (between daily mobility and interregional migration). Currently, this phenomenon is expanding among childless couples who choose not to share the same roof, but rather to meet to spend evenings, weekends and holidays together (De Singly 2000: 7-18). The main reasons for this practice include fear of boredom in the relationship and a desire to share only quality time;
- Short-term tourism (between daily mobility and travel). Leisure mobility that combines holidays and weekends have expanded exponentially in the last decade (Potier 1996). This mobility often takes the form of discovering a city and its cultural wealth in a relaxed manner.

There are undoubtedly many other examples. The common thread of these new forms of mobility is that they combine connexity, reversibility and ubiquity: they have consequential spatial impact characterised by so-called tunnel effects, and consist in abandoning the notion of irreversible mobility by developing forms of movement that allow people to avoid having to choose between alternatives such as moving or taking up a new job, between a partner and living alone, taking a vacation and travelling, and so on. They are also characterised by a form of ubiquity that consists in living several lives in parallel, in places that are physically far from one another.

Considering spatial mobility as a phenomenon likely to manifest itself in different ways, as I propose to do here, avoids the problem of focusing the attention of the researcher on the forms of mobility themselves, and instead focuses on the rhythms they make up when they are combined. Although overall those forms of mobility that are connex, reversible and ubiquitous are expanding, the rhythms born of the combination of forms of mobility are many and can easily vary greatly:

- It is not difficult to imagine the example of people who have highly developed residential and travel mobility, but a very strong residential anchor and no experience of migration. In this case, mobility is first and foremost reversible and based on short durations;
- We can also imagine people who, on the contrary, have highly developed residential mobility while their daily mobility very much relies on spatial proximity. In this case, we are faced with people who wish to restrict connex daily mobility, even if this means having to carry out more irreversible forms of mobility;
- We may also be faced with very mobile people who combine all the forms of spatial mobility, with notably a strong propensity to practise the new forms of mobility referred to above such as dual residences or short-term urban tourism. In this case, those concerned lead very fast-paced lives;

- Finally, we can imagine people who have lived through an experience of migration and numerous residential locations, but who travel little and have little daily mobility; their rhythms of mobility are then formed around long durations.

These different profiles of mobility are naturally linked to the positions of those involved, and all the cases outlined above take their meaning from the specific situation in which they occur[4]. Nevertheless, although each of the types of mobility identified maintains strong links with social position and culture, it must not be deduced that mobility is inflexible: *my entire conceptual approach supposes the existence of a degree of freedom*. Without this supposition, the interest of the construction that was just developed is very limited, given that the purpose is to discuss the possibility that social fluidification is linked to spatial mobility.

This postulate may seem banal, but in the area of spatial mobility it is not. The different forms of spatial mobility are still all too often interpreted unilaterally as the expression of power or of constraint. This attitude is often present among certain theorists such as Bauman (2000) when he evokes dominant mobiles as against dominated mobiles, but it is especially present in empirical research on migration, residential mobility and daily mobility. International mobility cannot, at first glance, be considered purely a function of the job market, nor individual habitat as the generalised reflection of family aspirations, the use of the automobile as a simple product of its efficiency. It is only the way in which people build these practices that can provide insight into what drives spatial mobility.

Supposing a degree of freedom a priori implies that mobility, or a lack of it, is the result of a choice made among alternative options. This supposition does not mean that I consider people as free a priori, but simply as actors in a situation. This choice of methodology does not make assumptions about the influence of social and spatial structures on the action that is precisely the object of the research. The question is not so much the existence of this degree of freedom as its extent and the variations of this extent as a function of factors of social differentiation. The debate surrounding social fluidification has to do with the extent of the degree of freedom and of its social 'localisation'.

From motility to mobility: towards a sociology of experience

I have ended up with two conceptual models: motility, which deals with people's potential, and mobility, which deals with their observable travel. What remains is to examine how motility is acquired and how it is transformed into mobility.

I have shown how access, skills, and appropriation come together to define motility with respect to the four main forms of mobility and the intermediary forms described above. Together, they constitute a propensity for mobility likely to be qualified in terms of the three oppositions developed in the preceding chapter, that is, contiguous-connected, reversible-irreversible, and unitary-ubiquitous. This is how, for example, someone can have motility strongly oriented towards contiguity, irreversibility, and unity, or on the contrary motility strongly focusing on connexity and reversibility, etc. Motility thus defines not only a propensity for mobility in terms of intensity, but also a propensity to realise certain forms of mobility instead of others, and to maintain a pace of life more or less oriented toward short temporal durations. Understanding how this propensity for mobility is constituted and how it is transformed into mobility means examining the intentions of those involved.

The acquisition of motility and its transformation into mobility takes place through decisions related to projects and behaviour that surpass spatial mobility alone. Finally, motility is at the service of people's aspirations, their projects and their lifestyles, and constitutes a 'mobilisable' capital for their realisation and their combination.

- With respect to the realisation of aspirations and projects and lifestyle, the acquisition of motility and its transformation into mobility is more often than not a means of carrying out a project or a consequence of carrying out such a project, and is rarely a goal in and of itself, with the exception of certain types of tourism, walking, or browsing.
- With respect to combination, aspirations are often contradictory, the projects sometimes irreconcilable, and so on. The transition from aspiration to project and from project to lifestyle implies often complicated juggling for which motility is mobilised. These choices increase in complexity with the number of people involved.

The supposition as to the existence of a degree of freedom, and the importance that I ascribe to logics of action and to people's projects place the analysis of the acquisition of motility and its transformation into mobility in a context of *sociology of experience* in the sense intended by François Dubet (1994), that is to say, 'a sociology of conduct that is dominated by the heterogeneous nature of their component principles, and by the activities of individuals who must shape the meaning of their practices within this heterogeneity' (Dubet 1994: 15).

The concept of experience defined this way is one of the rare paradigms of modern sociology that allows a theoretical project to be linked with empirical sociology of action. Built around the combination of logics of action, the notion of experience is characterised by three essential traits: (1) the heterogeneous nature of cultural and social principles that organise behaviour and which can come from instrumentality as well as from the integration of values, from habits becoming set, or from emotional elements; (2)

the all-important distance that individuals maintain with respect to one another and with respect to their practices and the opportunities they have, and (3) the absence of a central, apparently structuring principle in the construction of what is social (Dubet 1994: 16-19).

In this context, *the acquisition of motility and its transformation into mobility is built through the compromises made between aspirations, projects and lifestyle and is linked to multiple logics of action.* This point of departure places me before a multitude of cases that can be illustrated by the story of two women of the same age.

The first women aspires to a professional career in advertising. In addition to a recognised professional training course, she has acquired suitable motility by choosing to live in a city, having learnt two foreign languages and prepared herself to migrate to attain her career goal. She is married to an engineer who manages a consulting business that is well implanted locally and who is not likely to migrate, even temporarily. She is also attached to family life and plans with her husband to have children and to purchase an individual family home.

The second woman is divorced and has three children in her care. She works full-time as a social assistant. She is very busy and has little time for herself and for her private life in particular. She does not like to drive, but uses the car in her daily travel only because she judges that it is the sole means of transport that allows her to effect her schedule of activities. She has plans to move closer to her parents to facilitate taking care of the children, and would like to live in a larger apartment but does not have the financial means to do so.

In these two rather caricatured examples, neither the aspirations of the two women, nor their plans, nor their lifestyles are comparable. They base their decisions on different logics of action, and find themselves facing contrasting opportunities, and use motility and its transformation into mobility in a very specific manner. Nevertheless, beyond these differences, one variable unites them: they both have difficulty transforming their motility into mobility. The first uses motility as a strategic resource for a professional career and finds it hard to reconcile this aspiration with her plans to have a family. The second finds it difficult to transform her motility into mobility because of financial constraints and a complex schedule of activities that requires the use of a car. This brings me to the following proposal: *the degree of congruence between motility and mobility indicates the degree of fluidity of mobility practices.*

Applying the cases of these two women to all possible case scenarios shows that the degree of congruence between motility and mobility is always the expression of the more or less restrictive nature of the range of possible choices to which the person in question has access. To study the impact of spatial mobility on social fluidity, the multitude of possible cases of acquiring motility and transforming it into mobility can thus be summed up as the convergence between motility and mobility.

One aspect must nonetheless be stated clearly: motility does not necessarily have to be transformed into mobility. It can also be transformed into the use of means of telecommunication, with all that this implies given the importance of copresence for social insertion. But it is also quite possible to imagine people having skills or access to a broad spectrum of options, and choosing to realise certain forms of mobility but perhaps not many. In such a case, the potential for mobility, or motility, remains unexploited to a large extent, but this does not mean that there is divergence between motility and mobility; on the contrary, the degree of convergence between motility and mobility is measured on the basis of the experiences of those involved, of the meaning they give to their practices, and not by simply opposing motility and mobility.

Four research questions on fluidity

At the end of Chapter 2, I concluded that examining the fluidification of the social implies dissociating the potential for mobility from mobility itself, and dispensing with the idea that very mobile people have more freedom than their more sedentary counterparts, and finally that fluidification theses constitute a controversial interpretation of the evolution of current-day societies. We know that spatial mobility is becoming more and more reversible, more and more connex, and increasingly ubiquitous; we also know that the speed achieved by technological transportation and telecommunications systems is compressing time and space; what we do not know is whether these changes are accompanied by a greater degree of freedom in lifestyles. If this were the case, then differentiation would be a function only of the difference between people's aspirations and their projects.

It is these considerations and the critical examination of the concept of mobility that have prompted the proposal of a conceptual tool for measuring spatial mobility in all its complexity. This device allows the problem of the possible social fluidification produced by spatial mobility to be reformulated around four questions that relate to the degree of convergence between a person's motility and his or her mobility:

- Are people more free when they are more mobile? Or, *is the convergence of motility and mobility stronger among the most mobile people?*
- Do people seek maximum speed in their mobility? Or, in other words, *to what extent does motility converge with the search for maximum efficiency in the area of mobility?*
- Do all aspirations to mobility find fertile ground for their realisation? Or, put differently, *is it more difficult to transform certain types of motility into mobility?*

- Does speed potential produce fluidity? Or, *to what extent does the potential for speed facilitate the convergence of motility and mobility?*

I will examine each of these questions in the next four chapters on the basis of survey data. The analysis will be conducted from four bodies of data: a corpus of in-depth interviews on the different ways of organising daily journeys (survey by semi-directive interviews with 25 respondents in different Swiss cities); the data from a quantitative research study of the rationales that underlie modal practice (survey based on 3,000 people in Bern, Besançon, Geneva, Grenoble, Lausanne and Toulouse); the quantitative data of the research relating to the impact of the car on urban dynamics (enquiry based on 5,500 people from Île-de-France, Lyon, Strasbourg and Aix-en-Provence); case studies carried out on the co-ordination between urban development and public transport infrastructures in the framework of the European programme of research, COST 332 (case studies carried out in Switzerland).

These rich data will allow me to begin to analyse the use of speed potential provided by transportation systems. In this respect they are very useful for testing the notion of motility.

Notes

1   The term motility is used in biology and medicine to refer respectively to the capacity of an animal to move (such as the motility of a fish), of a cell or an organ (such as the eye). In sociology, it is not totally foreign, since Bauman uses it sporadically in 'Liquid Modernity' (2000) to describe the capacity to be mobile, and it is also found in sociological analyses of the body (Mol and Law 1999) to describe the body in motion.
2   This wording is a means of avoiding making reference to areolar limits.
3   An example is the seasonal migrations of English aristocrats at the beginning of the last century to the Côte d'Azur or to Switzerland.
4   Thus, the first profile is typical of families with children where both partners work and where there is a comfortable income, the second is quite characteristic of the life of a student living on campus, the third resembles the lifestyle of an executive with no children in the household, and the last brings to mind the mobility of an immigrant worker.

4   Mobile, therefore Free?

After three very abstract chapters, the time has come to test the theory on empirical data. The following four chapters will be devoted to the two goals of fine-tuning the concepts proposed and contributing to the debate on the social impact of time-space compression. Naturally, these four chapters will not allow for the stakes set forth here to be examined exhaustively, if only because the survey data on which they depend concerns relatively specific aspects insofar as it deals with the appropriation of speed potential offered by land transport systems. This data was retained because of its unique nature; this deliberate choice on my part is a reflection of the lack of in-depth research in the social sciences on transportation systems.

This chapter is devoted to the role played by spatial mobility in a person's life. Its objective is to show to what extent intensity and forms of mobility are linked to the broadness of the degree of freedom in the way a person lives. The idea is to measure whether convergence between motility and mobility is more advanced in people who are more mobile. This analysis will provide the basis for empirically testing the dual affirmation that to be more mobile means to have greater 'freedom', and that the development of connex, reversible and ubiquitous forms of mobility is the reflection of this presumed freedom.

The analysis is based on 25 interviews focusing on the role of spatial mobility in scheduling people's activities. The interviews were conducted in the four Swiss cities of Basel, Bern, Geneva and Lausanne among the users of central railway stations[1], and lasted from 45 to 60 minutes. They were conducted by the same person in German in the German-speaking area and in French in the French-speaking area, and then transcribed in to French. These interviews were part of research work involving railway station users, that also included a statistical analysis and a qualitative survey that will not be made use of here[2]. The very targeted sample allows for a detailed analysis of the aspects linked to mobility in the lives of public transport users who present very different characteristics in terms of gender, social and demographic traits, way of life, and family situations, since the interviewees included students living at home, retired couples and single people, as well as young couples with no children and families.

The analysis of this data will be broken down into three stages. I will first examine the different ways in which activities are planned, showing the role that mobility plays in these approaches. Next, I will show how ways of planning one's activities are constructed around specific motility types. Thirdly, I will measure the degree of convergence between motility and mobility, according to the frequency and diversity of the mobility realised.

Four ways of planning activities

The analysis of the way in which those interviewed plan their activities and projects distinguishes four ideal types in which mobility is used differently[3]. They are constructed around two elements: 1) the will to separate the sphere of activities belonging to social life[4] or, on the contrary, to integrate them; 2) the perception of travel times either as a duration or else as an intensity (in the words of Zarifian, 1999), that is to say, the attempt to plan the quantity of time allotted to different activities, or rather the attempt to have quality time in terms of its content. Table 4.1 illustrates these four ways of using mobility in one's life course.

Table 4.1   Mobility in a life course

	Extensive time	Intensive time
Distinction of the spheres of social life	Efficient alignment	Sensorial quality
Integration of the spheres of social life	Pre-planning	Open to opportunity

These four standard types can be described more precisely, while relating them to spatial mobility, in the following manner:

- **Efficient alignment.** This first way of organising one's tasks consists in planning a succession of activities whose location is pre-determined, and linking them by mobility slot. In this approach, travel is not usually used to carry out activities *en route*, but rather to provide the most effective possible link between places, and especially the home. For these users, the best feature of travel is its brevity, because travel is considered a waste of time, as for a 37 year-old married woman who said that she 'organises herself to travel the least and the most efficiently'. In this context, these time slots come in between other events. Efficient alignment means utilising speed potential that is provided by motorised means of transport. Often, rapidity at this point means being at home sooner, to 'have more time'.
- **Sensorial quality.** The second means of planning consists in carrying out one's activities by planning their succession by slots of mobility, but giving importance to the sensorial quality of the travel times. Although, as in the case of efficient alignment, daily motility does serve the purpose

of linking activities, in the case of sensorial quality efficiency is not the priority around which daily mobility is organised; here the primary goal is seeking pleasant experiences. Travel is no longer an in-between activity, but instead serves a purpose as leisure time between planned activities, such as 'getting some exercise or going for a walk' (a 30-year-old woman living alone). The sensorial quality consists *de facto* in seeking a residential location that has user-friendly spatial proximity and is rich in amenities, in daily life. This quality is formed around the use of proximity transportation means: walking and cycling.

- **Pre-planning.** This is a way of planning one's mobility so that travel is integrated with activities carried out en route. The time and space allotted to these activities is minutely **predetermined**, and the spheres of activity are mixed together, as in the case of a 57-year-old man living alone, who says 'I plan everything, from A to Z'; he believes he has 'an ounce of creativity with respect to short-term changes'. For example, he may use a train journey to relax or to work, or home may sometimes become his workplace, and so on. The organising principle is to seek the combination of time-space that allows the maximum of activities to be carried out. Mobility is even more of a formative element of daily mobility in this case: the ingeniousness apparent in the alignment of activities that make up daily mobility depends on the activities carried out, and the journeys are themselves often used to carry out activities, for example in railway stations and during a journey. The replies of the interviewees using this approach showed that pre-planning is built around the intensive use of public transport.
- **Flexible.** This fourth means of planning consists of combining the different spheres of activity while seeking to exploit them fully. In contrast to the pre-planning method, which places its minute and rigid schedule foremost, this method above all involves **creativity** in the planning of activities. These respondents, who generally travel in wide loops, describe their mobility in very reactive terms, and this approach is like a stimulus that generates opportunities to be seized, as in the case of the married woman in her late fifties and who commutes a long distance, and who considers that 'every minute can be utilised'; she takes advantage of opportunities to combine activities offered by each context to the maximum. The schedule of her activities can thus change at any moment; mobility in this case is at the heart of planning since it provides the opportunities around which the schedule is built. For example, the chance meeting of a friend can modify the schedule a little or a great deal. Being open to opportunity is associated with the combined use of means of transport.

The interviews clearly demonstrate that the four ideal types are explained by the complexity of the planned activities. In general, the scheduled activities are linked to the method used to plan them by the diversity of the activities and the spatial and time constraints.

Thus the respondents who can be characterised in terms of efficient align-ment have schedules that are rather simple and where their family situations mean that there are few mandatory activities (those who were interviewed were either students or retired). This search for efficiency is often a reaction to long travel times: the idea is to fight the barriers of time. This is notably the case of a student living at his parents' home, who must travel for one and a quarter hours by public transport instead of 45 minutes by car to reach univer-sity, and who systematically uses the car for this purpose whenever he can.

Those interviewed who were seeking sensorial quality have schedules that are more or less complex according to each individual case, but with importance given to mandatory tasks such as work or housework. This is often the case of those who work a great deal. The search for quality in travel as part of daily life is a reaction to these obligations, and allows people a moment of respite when they can savour the present before beginning other activities. This is the case of a young married woman, who has two young children and works, and who chose her neighbourhood because it is a 'vil-lage within the city' with many small shops where customers are greeted by name, and which represent familiar meeting places where she feels good.

Pre-planners often have numerous activities that are mandatory; they often have family responsibilities. It is not uncommon for this complex schedule to be further complicated by pronounced spatial constraints, such as a job far from home, intricate timetables for ferrying children to activities, and so on; finding solutions requires a good deal of mental agility. Planning every minute of each day is therefore a must in order to avoid wasting time and to achieve all the planned activities.

Finally, we come to those respondents who were open to opportunity: although their scheduled activities were either simple or complex, they are characterised by the flexibility of their timetables (flexible working hours, 'floating' meeting times adjusted at the last minute using a portable tele-phone), which gives the impression of having time. This is the case of a mar-ried woman who has a pendular travel route to her long-distance workplace, and who says she changes her planned activities according to her mood. Being open to opportunity implies having this flexibility, without which it would be impossible to respond to contextual stimulation, even for an activity that does not take up much time. During the interviews, it became apparent that open-ness to opportunity is sometimes linked to a strong professional commitment.

Motility

These four ways of planning activities require specific motility in terms of access, skills, and appropriation. Two initial aspects can be identified.

1. First of all, there is the reasoning behind the choice of motility. Those who opt for efficient alignment of activities develop very good knowledge of available transport, its strengths and its weak spots. They become experts in which train is the best to catch, which route is quickest. Any possible speed potential is evaluated according to the efficiency of the trip it allows, as in the case of a 20-year-old student who lives at home with her parents and who has a custom-designed, pocket-size timetable showing all the quickest public transport transfers and the most rapid route she can take to get home from university at a given time.

   Those who seek sensorial quality are very well-versed in the composition of the city around their homes and are seeking spatial unity in everyday life (especially proximity to the workplace). Their evaluation of a context focuses on the quality of life it offers, of how easy it is to travel by foot or bicycle, how comfortable the public transport network is, the degree of tranquility, perceived pollution and available urban facilities that exist. These people prefer urban centrally-located places, as is the case of a young married woman interviewed, who does not work, has a young child to look after, and who appreciates 'being in the city centre, where getting around on foot for leisure activities in the evening, for example, is easy'.

   Pre-planners who are open to opportunity are very much aware of transport access and seek to live close to transfer points. They are very knowledgeable about the rules of the game regarding transportation network use, (schedules, lines, how full the vehicles are, parking facilities, traffic conditions) and as to what is accessible from where. These skills constitute a *sine qua non* condition for their scheduled activities; speed potential is judged according to what it provides access to in terms of equipment and services, and not in terms of pure rapidity.

   This first aspect shows that in the four standard types, the planning of activities is not based on the same reasoning, nor on the same point of view. Indeed, as far as the viewpoint is concerned, planning either starts from the ground and then takes account of the available networks, as is the case of those pursuing efficient alignment of activities who seek the fastest way of linking predefined points of origin with their destinations, or else it departs from the networks and then takes into account the ground to be covered, as in the case of those seeking sensorial quality who evaluate the quality of nearby trips so that they can be more sedentary as a result; still a third possibility is that the point of departure of this planning is the correspondence between the network and the ground to be covered, with a search for a residential location that provides access, via the networks, to ground that is spatially as far away as possible, as in the case of the different pre-planners and those open to opportunity.

2. The second aspect at work in the motility of the different standard types is the propensity to carry out different forms of spatial mobility and the means used to realise them.

Those attached to efficient alignment have a fairly weak propensity for spatial mobility in general. In their everyday mobility, they tend very much to use public transport. They do not have a car for financial reasons or for reasons of physical inaptitude or conflicts in the allocation of their resources.

Pre-planners have a propensity for daily mobility and for travel involving the highest degree of connexity. Take for example the case of a 51 year-old married man with two young children, who has a long-distance, pendular work route, who explains the fact that he has not moved to the city where he works because 'the children's grandmothers live in Geneva, which is practical for babysitting'. These respondents have automobiles, but use public transport for their longer journeys so that they can use their travel times.

Those sensitive to sensorial quality have a very strong propensity to develop micro-mobility on a daily basis , based on spatial proximity combined with a high propensity to travel, to residential mobility, and even to interregional migration.

As for those open to opportunity, they have a propensity to carry out a wide range of mobility forms with a pronounced preference for the new intermediary types outlined in Chapter 3. 'It would not be uncharacteristic for me to leave for New York tomorrow if I found an inexpensive ticket offer today' says one single retired man.

Thus, motility is oriented towards an entire range of specific forms of mobility that is more or less broad according to each individual case. In this context, it is important to note that it is among the two standard types characterised by the search for intense time that we find the respondents most likely to engage in the widest range of forms of mobility (sensorial quality and open to opportunity). It remains to be seen whether this more varied propensity is the reflection of having more room to manoeuvre within a life course.

The degree of congruence between mobility and motility

What types of mobility are associated with the four standard types identified? So far, I have concentrated on the description of these standard types and on the motility to which they are each linked. This is the heart of the chapter's focus: I have just highlighted the fact that the standard types are characterised by propensities to engage in the different forms of mobility that are more or less varied and wide-ranging; to what extent is there congruence between these forms of motility and the spatial mobility that is effectively realised?

Divergences in each standard type

An important first element to note is that in the four standard types the convergence between motility and mobility is variable. Thus, in each of these types there are people who are more or less satisfied with their motility and their mobility.

One of the reasons for dissatisfaction among those attracted to efficient alignment is that they deem that this alignment is not efficient enough. This discontent often comes from not having a car. Thus, the desire to have a car fully at one's disposal is sometimes present in those who swear by efficient alignment; this is true of the student living at his parents' home and who must make very long journeys on public transport to get to university. Similarly, some respondents seeking sensorial quality are not satisfied with their environment and constantly experience the desire to move. An example of this is one young man living alone who says he 'experiences moments of stress on his bicycle in traffic every day' and who is seeking to move to a quieter neighbourhood. Pre-planners sometimes suffer from transportation that is not sufficiently reliable, and which puts them under pressure in their mobility and forces them to abandon some of their activities or to postpone them. Dissatisfaction is also present among those open to opportunity since they often consider that the opportunities provided by their context are insufficiently diverse, and do not allow them to be sufficiently mobile and to combat the routine of daily life.

Dissatisfaction sometimes comes from an erroneous evaluation of the conditions of mobility. During the interviews it became especially evident that this was the case for residential mobility. In one case in particular, a mistaken evaluation of the public transport available in the neighbourhood and of the related facilities means that one family has difficulty getting around. Another example involving the young man living alone and seeking sensorial quality involved his abandoning the bicycle in favour of public transport to go to work because of the traffic; he had miscalculated the conditions for travelling by bicycle around his neighbourhood.

Finally, the divergence between motility and mobility is sometimes the result of family compromises. Some respondents seeking efficient alignment have been forced to move, such as the 37 year-old married woman who, in 'following her husband', found herself in an unfavourable context for using public transport when she has no intention of driving a car. There is also the case of the pre-planner who gave up on the idea of moving into an individual house with his family because of the spatial-time tension that such a move would cause in his scheduled activities.

Thus, in each planning method, the congruence between motility and mobility varies. Some people are restricted by neither their motility nor their mobility; for others, in contrast, motility converges only partially with mobility, thereby provoking tensions that produce dissatisfaction:

•   Either motility is restricted by one of its components – access (such as the lack of availability of a car), skills (not having a driving licence), or appropriation (mistaken evaluation of available options in a given context), or

- There is incompatibility between motility and mobility, in which case the realised forms of spatial mobility or of being assigned to a sedentary state do not correspond to the desires of the respondents.

The congruence associated with the intensity of mobility

I therefore note that the respondents with a propensity for the most diversified forms of mobility are not necessarily those who are the most satisfied with their motility and/or their mobility.

If the degree of freedom cannot be linked to the methods of planning activities, then can it be linked to the intensity of mobility? Do those interviewed who travel the most see a greater degree of congruence between their motility and their mobility?

If the intensity of mobility is to be measured by the diversity of the forms realised and the frequency of the movements, let us immediately recall that the most mobile people are found among the respondents seeking sensorial quality and those open to opportunity. This is hardly surprising since these two standard types are those characterised by the propensity to engage in the most diverse forms of mobility.

Secondly, let us point out that the most mobile people are characterised by a particularly high degree of convergence between their motility and their mobility. The two examples which follow illustrate this perfectly.

The first case is a 'sensorial quality' example involving a young woman working in the hotel industry in Basel. Originally from the Geneva region, she has worked over a ten year period in several German and Swiss cities; she thus has experienced migration. Her friend, who also works in the same business, lives in Zurich and the young woman often spends the weekend at his home, thereby realising intermediary forms of mobility. Moreover, she enjoys taking long trips and does so regularly. Her workplace is located close to her home in the city centre, which is particularly practical in that she works irregular hours. She enjoys walking in her neighbourhood as a break from her often busy schedule, and is very sensitive to the sensorial quality of urban spaces. She has no car by choice.

The second example involves a female executive in her fifties who is 'open to opportunities' and lives near the city of Lausanne. She travels a pendular route to work, and moved with her husband, also an executive and long-distance commuter, to be closer to German-speaking Switzerland where her husband often travels on business. She has one son who lives abroad. The couple have quite a developed programme of mobility for evening and weekend leisure activities and effect intermediary forms of mobility. She works four days per week and spends four hours travelling by public transport each day, including two hours by train. She chose this latter means of transport because she considers train travel as free time and is open to opportunities that come up, especially via her mobile telephone. She has a car entirely at her disposal.

In both cases, motility is optimised: the transport used was chosen, the skills match the lifestyle and the appropriation of the context is very open and complete. Although all the potential constituted by the motility is not transformed into mobility (some skills and some available transport are not utilised), the mobility that is realised is totally congruent with the desires of these two women, who both describe themselves as being very free.

And yet, when one looks more closely at the situation, this result is entirely ambivalent for two reasons.

Those who are the most mobile attribute a great deal of importance to their careers. The mobility of employees is often a reaction to the demands that employers make of their executives, since arming oneself with a wide range of choices with respect to mobility is to a certain extent an obligation for individuals who wish to pursue a career. Moreover, considering that the schedules of activities of such individuals are blocked by their careers, making mobility converge with motility is the only degree of freedom they have. The search for sensorial quality in everyday travel makes sense in light of schedules that are quite restrictive. It is a way of taking a breather – unlike in the first case above – and of enjoying a moment of freedom in everyday life. Openness to opportunity is in many respects a state of mind that allows the person to regain access to a little freedom in a complicated situation where different domains of activity mix.

The apparent 'freedom' of the most mobile respondents is a result of compromises in their life courses. The degree of convergence between motility and mobility is often the result of the priorities of those interviewed and the process of allocation of their resources in accordance with these priorities: some people arm themselves with 'freedom' in their mobility while others use the room to manoeuvre at their disposal differently. The 'sensorial quality' and 'open to opportunity' respondents, who are characterised by a strong degree of convergence between their motility and their mobility, use the room to manoeuvre they have for both their motility and their mobility. This does not mean that those who make use of their room to manœuvre differently are less free. The students seeking efficient alignment of their travel could have another type of daily mobility; for this they would have to move or purchase a car, which would mean paying either rent or maintaining a vehicle. They would have to work in order to obtain this money and would thus be forced to have a more complex and more restrictive schedule of activities. Instead, they use their mobility conditions to have free time and to allocate their resources to other projects (going out, travelling, etc.). For pre-planners, reconciling numerous activities is often the expression of a freedom that is paid for by the energy spent planning and by a significant allocation of financial resources to transportation.

Conclusion

The results of the analyses show, via the four standard types presented, four very different relations to territory. They consist of juggling mobility and sedentarity and they form specific space-time rhythms.

These four relationships to the ground covered when travelling can be placed in the context of the 'structured-confined' – 'unstructured-infinite' dichotomies inspired by the work of Deleuze and Guattari mentioned in Chapter 1. At one end of the scale, sensorial quality is clearly related to structured mobility by restricted territory or ground, while at the other end openness to opportunity is related to a form of mobility that produces infinite territory to cover. In the first case, mobility serves as a link between places with which domains of activity delineated by areolar barriers are associated. In the second case, the person moves along with his or her territory and mixes domains of activity as and when the opportunity presents itself, ignoring the functional specificity of each space. The other two standard types are located half-way between these two extremes; those people who seek to align their activities efficiently move within territory that is confined, but which leaves them free time. Pre-planners move within territory that is highly structured, but which they appropriate by mixing different domains of activity.

The conclusions at which I have arrived are directly linked to this observation; they can be formulated using two affirmations that restate the research questions asked at the beginning of this chapter:

Nothing shows that the most spatially mobile people have more freedom in the way they conduct their lives. The most mobile respondents are those who have the greatest degree of congruence between motility and mobility. It cannot be deduced that they have a greater degree of freedom in the way that they conduct their lives in general. A 'freer' mobility is often the sign of people having assigned the degree of freedom that they have to their mobility rather than to something else. There is thus ambivalence: using the potential rapidity offered by technological transport and telecommunications systems broadens the potential in terms of mobility, but this broadening is used to reconcile more constraints rather than to obtain more freedom. It is as if this potential gave new freedom to those respondents whose lifestyles are very much confined by spatial restrictions. Mobility gives new freedom to those people who would not otherwise have any.

Spatial mobility does not necessarily equal social fluidity. The most mobile respondents were found in two types that match the models of structured-confined and unstructured-infinite. It can thus be deduced that there is not necessarily a correlation between social fluidity and spatial mobility: people can be very mobile without having a lifestyle based on fluidity, but one that is based instead on the will to separate their lives socially and spatially into distinct areas. Those respondents belonging to the standard type that I called 'sensorial quality' have a strong propensity to effect uninterrupted, irreversible and unity-seeking mobility. Although they are sometimes very mobile and possess considerable purchasing power, they do not evolve in a fluid social world because their mobility respects the barriers of areolar territories such as the neighbourhood as a unifying space of daily life, or it respects the spatial differentiation of spheres of activity in general. The interviews furthermore highlighted the fact that the most mobile people place work at the centre of their existence. Their high degree of mobility is often a more or less direct reaction to the flexibility required of them by their companies in their capacity as executives. Their mobility therefore has more to do with their submission to structures rather than their escaping from them.

Notes

1   All the interviews and their transcription were carried out by Michael Flamm.

2   This research was summarised in a report entitled '*Entre rupture et activités: vivre les lieux du transport – De la sociologie des usages à l'aménagement des interfaces*' (Kaufmann, Jemelin and Joye 2000) and in summary reports published in French and German.

3   I attribute the interviewees' different ways of planning activities to one of the ideal types according to the degree to which they resemble these types, and even in some cases to more than one type if they use a combination of types in carrying out different activities.

4   In this chapter I will make use of the distinction proposed in Kaufmann (2000) between four spheres of social life: the workplace, the home, free time and citizenship. These four spheres differ from one another in nature as either free or forced, and whether or not they are remunerated.

5 The Use of Speed Potentials

I have extensively criticised the confusion between the potential for mobility made possible by transportation systems and spatial mobility itself. The time has now come to discuss this point empirically by means of a simple question: to what extent do actors attach importance to speed above other factors?

The growth of connexity as well as the reversibilisation of spatial mobility and its ubiquitous nature have led many authors to believe that the search for speed is the motivation behind spatial mobility practices. This is the same as saying that there is some sort of automatic use of transportation systems that are faster. In the same vein, according to the latest analyses social fluidification is the product of technological advances that allow speed to be used to cover the longest possible distances.

In many ways the present chapter tests this line of argument. The ideal types that are developed in it shows that the most mobile respondents are not necessarily those with the most developed forms of connex, reversible and ubiquitous mobility, and that they do not necessarily move in a world of greater social fluidity. Thus these ideal types suggest that the actors do not necessarily have as their objective maximum speed nor the desire to escape the boundaries of social and territorial structures.

In this chapter, I will study the problem from the point of view of a very specific question: what are the motivations for the daily modal practice of those who have both an automobile at their disposal and good quality public transport in the vicinity of their homes? The data on which the analysis will be based comes from a comparative study in France and Switzerland of six cities with contrasting speed potentials for both public transport and automobiles due to their existing networks and spatial morphologies.

The methodology used consists in a telephone survey of people who are theoretically in a position to choose between using public and private transport; they are individuals with access to their own vehicles and who have a direct, primary urban transport network route linking their homes with the centre[1]. The advantage of this procedure is that it allows me to focus the analysis on a specific part of the population, those representing the principle target of policies aimed at creating a modal transfer. It also allows me to compare the daily commuting habits of these respondents in the six urban centres studied.

This approach also presents the advantage of comparing contrasting urban phenomena in two distinct political systems (French and Swiss) and cultures (francophone and germanophone), thereby enriching the argument still further. The six urban centres discussed – Besançon, Grenoble and Toulouse in France and Bern, Geneva and Lausanne in Switzerland[2] – are characterised by the contrasting rates at which their different means of transport are used[3]. Cultural differences are often put forward to explain this state of affairs: the Swiss, for example, are said to be more environmentally aware and therefore more inclined to use public transport than the French (Ziegler 1995: 38-43). Similarly, the French and Swiss are said to have rather different attitudes towards the law. These generalisations never having been verified, however, they are really no more than national stereotypes.

Available transport and usage

In this research, I was able to observe a similar contrast in terms of spatial morphology (Table 5.1), spatial-temporal uniformity in the public transport system (Table 5.2) and available parking spaces in the city centres (Table 5.3).

Table 5.1   Spatial morphology indicators of the urban centres

	France			Switzerland		
	Besançon	Grenoble	Toulouse	Bern	Geneva	Lausanne
Total population of the urban district (1990-1991)[4]	123,000	405,000	650,000	332,000	424,000	295,000
Proportion of the population residing in the city centre	93%	37%	25%	41%	40%	43%
Density of the population in the city centre (per hectare)	17.4	82.7	80.8	26.4	107.4	30.1
Proportion of jobs based in the city centre (1990-1991)	94%	46%	33%	69%	56%	55%

Note: The urban district of Toulouse being very wide, we have included the perimeter of the 'centre étendu'.

In spatial terms Besançon is a small, predominantly mono-centred city (in the sense of a strong city-centre without surrounding suburbs). It is the centre of a region where there are very few towns. In Grenoble, despite the centre being dense in terms of businesses and homes, there is also a great deal of commerce located on the outskirts. The spatial structure, therefore, encourages journeys to and from the suburbs. In Toulouse, which is the biggest of the cities studied, the suburbs are highly developed. They attract a large proportion of the big employers and many out-of-town shopping complexes. Bern, although very spread out, is distinctly mono-centred, both in terms of employment and shops. For more than three decades the planning of the area has been based around the accessibility of public transport. In Geneva the centre is very dense both in terms of housing, employment and shops. However, the expansion of out-of-town shopping complexes is likely to lead to daily centrifugal movement. Lausanne is a less dense urban centre, characterised by a vigorous commercial development around the southwest approach to the city.

On a public transport level, a comparative study of the number of passengers per day on the principal network lines brings to light time-space coverage which is very varied, to say the least. In Grenoble and Toulouse the bus service is two to four times less frequent than that of the metro/tramway system, in contrast with the Swiss towns of Bern and Geneva where the qualitative uniformity of all services is guaranteed. Moreover, the availability of public transport on Saturdays and Sundays is very distinctly reduced in the French urban districts.

Table 5.2   Number of daily service trips on the principal urban transport networks[5]

	France					Switzerland					
	Besançon	Grenoble		Toulouse		Bern		Geneva			Lausanne
	Bus	Tram	Bus	Metro	Bus	Tram	Bus	Tram	Bus	Metro	Bus
Mon-Fri	80	170	70	190	100	170	160	170	140	130	130
Sat	60	120	60	180	80	150	130	150	130	120	80
Sun	30	90	30	120	30	130	120	130	80	100	60

As regards town centre parking, I was able to observe considerable differences in the extent of available parking spaces (Table 5.3); Bern has a mere 3,800 places, while Toulouse boasts more than 29,000. Of course these figures only make sense if they are related to the population of the urban centre. Based on this criterion, it is Besançon that has the most well developed central parking facilities, followed by Grenoble and Toulouse. In Geneva and Lausanne this ratio, based on the available parking spaces linked to the population, proves to be much lower than in the French urban centres. In Bern this ratio is the lowest – ten times lower than in Besançon.

Table 5.3   Available parking in the city centres

	France			Switzerland		
	Besançon	Grenoble	Toulouse	Bern	Geneva	Lausanne
Public parking	58%	62%	83%	66%	75%	64%
Private parking	42%	38%	17%	34%	25%	36%
Overall total	12.000	21.000	29.900	3.800	11.400	10.800
Ratio spaces/inhabitant	0.10	0.05	0.05	0.01	0.03	0.04

The analysis of the modal habits of the people interviewed, who, it must be recalled, are those who have access to their own vehicles as well as to public transport near their homes, reveals two clear tendencies:

- **A substantial difference in all cities between the frequency of use of the car and of public transport** In the six urban centres studied the car is very often used more than once a week, while public transport is only used very infrequently. When someone has a private vehicle at their disposal and good quality public transport near their home they generally use the car far more. The majority of people who are theoretically in a position of 'modal choice' tend to work out their activity schedules around the use of the car rather than public transport (see Table 5.4).

Table 5.4   Modal habits of drivers/public transport users in a theoretical situation of 'modal choice' [n=3001, 500 per city]

Private vehicle	France			Switzerland		
	Besançon	Grenoble	Toulouse	Bern	Geneva	Lausanne
< than 2-3 times per wk	85%	96%	90%	83%	89%	92%
< than 2-3 times per wk	13%	2%	8%	15%	10%	7%
Never	2%	2%	2%	2%	1%	1%
Public transport	Besançon	Grenoble	Toulouse	Bern	Geneva	Lausanne
> than 2-3 times per wk	26%	27%	25%	60%	29%	29%
< than 2-3 times per wk	38%	40%	40%	35%	51%	51%
Never	36%	33%	35%	5%	20%	20%

• **Differences between the various urban districts in terms of frequency of usage of public transport** Bern is the city where public transport is most used, followed by the two French-speaking Swiss urban districts (Geneva and Lausanne) and the three French ones (Besançon, Grenoble and Toulouse). These findings, based on a sub-population, confirm the International Public Transport Union's overall annual results and the results from the official French and Swiss household surveys. The most interesting element to come out of the findings is the highly contrasting rate of total non-usage of public transport. In Bern only 5% of the respondents never use public transport. In Geneva and Lausanne this figure rises to 20%. In the French urban districts studied the figure is more than 30%. Thus, in the five French-speaking urban districts, between a third and a fifth of the people in possession of their own vehicle, with high quality public transport close to their homes, never use public transport.

How can these two trends be explained? Even if they coincide a priori with the spatial structure, the quality of public transport and the amount of parking spaces available in the centres, they still do not reveal anything about the rationales at work in the different urban contexts. In particular, they reveal nothing about whether the actors are seeking to minimise their travel times or instead base their modal practice on other criteria. To what extent do the contrasting modal habits observed reflect the compared efficiency of the automobile and that of public transport in the six cities studied?

Asking the question in this manner brings us to focus on the appropriation dimension of motility. The motivations underlying modal practice are the product of the way that the actors appropriate the context in which they find themselves. Based on these rationales, I will develop a typology that links them to the use of the automobile and public transport (and thus to mobility). This typology will notably pave the way for discussing the meaning given to access (we may recall that all the respondents questioned have access to both a car and to public transport) and the way in which it is used by actors.

Three rationales at work

Three rationales stand out as central in explaining modal practice: a comparison of journey times, cultural predisposition to the use of different modes of transport and the anchoring of different modal habits in the way people live their lives.

Comparison of travel times

The first rationale that stands out as preponderant in this 'modal choice' is the comparison of journey times by different modes of transport. When the car is quicker than public transport to get someone to work it is used by 81% of people, while if the situation is reversed, only 43% resort to the car (Table 5.5). The simple comparison of journey times, however, cannot explain all the forms of modal practice.

Table 5.5   Journey times and modal practice

n = 1252	Public transport faster	Duration comparable	Motor car faster
Motorcar	43%	63%	81%
Public transport	57%	37%	19%
Total	100%	100%	100%
Row %	10%	29%	61%

Note: This analysis has only dealt with the users of private cars and public transport, and has excluded other modes of transport. Moreover, it has only concerned itself with the urban districts of Bern, Geneva, Grenoble and Lausanne.

Although there is a link between comparisons of journey times by car and public transport and modal practice it by no means constitutes an automatic causal relationship, as the non-symmetrical nature of the relationship between modal practice and time comparisons suggests[6] (Table 5.5) and several results obtained from other sources confirm.

For journeys to and from work, the parking conditions at the place of work affect whether the car is quicker or slower than public transport. This disguises the fact that, even when there is allocated parking at work, the decision to use the car is often not linked to any comparisons of journey times.

A proportion of the respondents tend to use the car regardless of the quality of the available public transport. To understand the practice of these respondents, therefore, comparing car and public transport journey times makes no sense. One could extend this observation to the choice of areas to live. Numerous choices in this regard are based on access to the road networks, without giving a thought to public transport services. These respondents, then, are impervious to what is on offer in this respect, even when they live close to a good-quality public transport route.

The perception of the quality of time spent doing something restricts rationale in terms of evaluating actual duration. The biases in the perception of journey times, often rooted in processes of self-justification of their own actions, mean that practically everybody who uses a particular means of transport for a given journey considers that they are minimising their journey time.

Taking all these results into consideration, how can one explain the relationship, despite everything observed, between the comparative speed of different means of transport and modal practice? It appears that this connection is a reflection of the fact that the motorcar is often faster than public transport. One must not, however, confuse this circumstance with sound reasoning behind modal practice; when other means of public transport are quicker than the car they are not necessarily used. This confusion between a fact (the car is often a quicker way of getting to a specific destination) and an explanation (the car is used because it is quicker) provides us with an explanation for why massive investment in the construction of public transport infrastructure is not generally rewarded by modal shifts away from the car.

Preference for use of the automobile

A second rationale relates to the social perceptions of different means of transport. A large majority of people interviewed prefer to use the car rather than public transport. This observation could be interpreted as an expression of a triad of very western values: speed, individualism (car journeys are undertaken alone or with 'chosen' passengers) and privatisation (car journeys are undertaken in a private space, totally under the driver's control). In addition to these 'qualities', it is also manifestly a symbol of liberty. On the other hand, all forms of public transport are defined by their contrast to these characteristics. They offer neither the possibility of traveling in private nor the individuality associated with the car. They also restrict their users to the constraints of set routes and timetables.

The above findings emerge from a corpus of adjectives used by respondents to describe the car and public transport. This method was employed in order to discover the social perceptions of the car[7], and produced some very contrasting descriptions. The car is linked to terms such as 'practical', 'fast', 'comfortable' and 'allows independence', while public transport, although qualified as practical, is also associated with such descriptions as 'slow', 'restrictive' and 'overcrowded'. The dominant social perceptions emerging from these descriptions concern the quality of the travelling time. The often virulent criticism of the car in the Germanic world hardly seems to scratch the surface of this social perception. Bern is no exception in this respect. This finding is a key point to reflect on when considering the differences between Swiss and French, Germanic and Francophone contexts: the dominant perception of the motorcar remains the same.

The order in which these adjectives are given when qualifying public transport, on the other hand, differs from one urban centre to another. Thus, the terms judged pertinent to describe public transport are, overall, similar in all the urban contexts studied, but their weighting is variable. Relating the body of adjectives used in each urban centre with the available public transport reveals a relationship between the quality of the public transport and the use of the adjectives 'restrictive', 'slow' and 'overcrowded'. 'Restrictive' is used more by the respondents living in urban districts where the nature of the network dictates more frequent changes; the term 'slow' is associated with the speed of the service as well as the structure of the network, while 'over-crowded' turns out to be linked to the most overloaded services. Moreover, the term 'ecological' is mentioned far more often in Bern than in the other urban districts to describe public transport, attesting to a greater awareness of environmental issues in this city. This is the only important difference noted between all the urban districts with respect to the descriptions used.

In terms of explaining modal practices, the preference for use of the car is manifested in the fact that public transport is usually only used when conditions for car use are unfavourable. The quality of the public transport system barely affects the decision.

Modal practice rooted in lifestyles

A third rationale refers to the embedding of modal choice within everyday practices. The result of this strong link is that modal practices are not inter-changeable, due to the fact that each means of transport defines the available combinations of activities in space and time. Thus, for example, the generally radial structure of public transport means that its use generally increases the opportunity to carry out various activities in the town centre. On the other hand, use of the car often allow drivers to take advantage of the commercial facilities on the outskirts of the town or city, road access for these places being almost always excellent.

The embedding of modal practice in lifestyles means that it is very difficult to modify. Any change has implications far beyond the sole domain of modes of transport. The extent of the cost associated with modal shifts towards public transport depends on government action in the area of urban planning and transport. In the urban districts where development has traditionally revolved around the motorcar, as is the case in Toulouse, this cost is distinctly higher than in the urban centres where growth has been strategically linked to bus and train stations. In this second scenario, characteristic of Bern, the opportunities to reconstruct spatial habits around the use of public transport are quantitatively more numerous and qualitatively more varied.

It is evident from our data that this aspect is a considerable obstacle to modal transfer. The exclusive users of the car consider that public transport is very inefficient and have a particularly negative perception of it. This perception stems directly from the spatiality of their activity schedules. If they used public transport they might often not manage to complete their activities, giving rise to a particularly critical vision of this means of travelling. They omit in their evaluation the fact that if they were users of public transport their schedules would have different space-time characteristics.

Between exclusive automobility and receptivity to other options

The three rationales that have been briefly described above tend to combine with each other and do not affect all the respondents in the same way. Likewise, their respective importance also potentially depends on the urban district under consideration. In order to study these aspects I have resorted to a typology. Based on modal habits, on social perceptions of the motorcar and of public transport and on the conditions of their use, this typology includes four types, each fulfilling a combination of specific rationales[8]:

- The first type, which we have named exclusive motorists, is composed of individuals who never use public transportation, even though they have high-quality services in close proximity to their homes. In fact, the use of public transport for these respondents tends to be effectively outside their realm of possibility, given their particularly negative view of it. Fuelled by the dominant perceptions we have spoken about and amplified by their lack of experience of public transport, their prejudices against it are great. Mainly comprising men of a high socio-professional standing, and living and working away from the town centre, this first group is characterised by a marked tendency to choose their non-constrained daily destinations *according to the perceived ease with which they can use their cars*[9]. Among these individuals a decision to frequent a particular place will depend to a large degree on its perceived accessibility by car.
- The second type, named civic ecologists, is composed of people whose values essentially revolve around respect for the environment. These individuals, therefore, vigorously distance themselves from the dominant social perceptions of the car and public transport, turning them around by stressing the disadvantages of the car and the advantages of public transport in environmental terms. They favour the use of public transport over the motorcar whenever it proves possible without wasting time or causing undue inconvenience. This type is predominantly composed of young people and women as well as respondents who work in the town centre. The logic which underlies the modal practices of these people harks back to Max Weber's concept of 'Wertrationalität' (1922): *the use of public transport stems more from a value system with which the person wants to adhere than from the quality of the transport on offer.* These 'civic ecologists' frequent the town or city centre a great deal, only rarely travelling there by car.
- The third type, which I term motorists compelled to use public transport is composed of people who adhere to the dominant social perception of the car and public transport. In a theoretical situation of modal choice they always prefer to use the motorcar and do not contemplate the use of public transport at all except when the use of their own vehicles is problematic.

All the individuals who make up this category are subject in varying degrees to such a constraint and are therefore users of public transport[10]. They arc set apart from the first type in two essential ways: firstly in that the use of public transport is not outside their sphere of possibility, being, as they are, regular public transport users. Secondly, they do not base their choice of shopping venues (city centre versus large out-of-town shopping complexes) around their accessibility by car. In the main this group consists of women and individuals of low socio-professional status, very seldom in possession of their own reserved parking place at work. To a very large extent unaffected by the quality of the available public transport, *these individuals will use the car each time that traffic and parking conditions permit it, and will not revert to public transport except when the opposite is true.* When confronted with parking restrictions these people have a clear tendency to modify their modal practice rather than their destinations.

- The last type, which I have named sensitive to time, is composed of people who base their modal practice on a comparison of the different journey times, cost, effort, convenience and other factors. These people are largely unaffected by the symbolic perceptions of both public transport and the car, having experienced both of them. Of all the people questioned in a theoretical position of 'modal choice', these are the only ones who consider themselves to be effectively in a position of choice between two alternatives. In terms of sociological profile, we note that 'time-sensitive users' can be found in all social categories. The rationale which underlies the modal practice of type 4 *results from a comparison of all possible means of transport, which, in turn, results in the choice of transport which allows them to get to their destination in the most efficient way possible.* Like the 'motorists compelled to use public transport', if they are confronted with parking difficulties they will undertake a 'modal transfer' rather than rethink their destinations.

These four types constitute specific appropriations of availability. They require skills and structure the accessibility to both the automobile and public transport that all the respondents benefit from. The four types are also characterised by their varying degrees of convergence between motility and modal practice. The exclusive motorists appropriate the context according to automobile access and develop the appropriate skills such as in-depth knowledge of the parking conditions on such a day at such a time, and so on, around the different destinations of their daily movements. Their mobility converges with their motility: while they often have good-quality public transport near their homes, they do not make use of it at all, no matter how good the service is. The civic ecologists appropriate their context by means of their convictions, and they ensure they have good public transport access and the option of walking or cycling, as well as sound knowledge of the public transport network and its potential. Their mobility often does not converge totally with their motility, as is proven by the frequent use they often make of their cars. The motorists compelled to use public transport have motility that corresponds quite closely to that of exclusive motorists, but they have a more favourable opinion of public transport and resign themselves to using it when conditions dictate that access by car to a certain place will be difficult. They have divergent motility and mobility in the sense that the use of public transport is a constraint for them. As for those 'sensitive to time', they have motility that is characterised by the desire to have access to both the automobile and public transport and to evaluate available transport according to the comparative rapidity of the various means. This evaluation requires specific skills such as attentive studying of timetables and very good knowledge of parking and traffic conditions. The mobility of the time-sensitive generally matches their motility; for example, for home-to-work travel, I observed that they effectively minimised their times[11].

The examination of 'weightings' of different types within the sample questioned brings to light considerable differences between the six urban districts (see Table 5.6). The most striking is just how variable the proportion of 'exclusive motorists' (type 1) is; it runs at 5% in Bern to around 20% in the urban centres of Geneva and Lausanne and at more than 30% in the French cities. This contrasting weighting co-varies with type 4, 'time-sensitive users' whose weighting is substantially larger in Bern than in other urban centres. We note that 'civic ecologists' (type 2) are in the minority everywhere, except in Bern where they actually represent 14% of the people questioned.

These differences show to what extent context affects the appropriation of the means of transport by the actors, on two levels:

- The different contexts are marked by specific values and a local political culture that have an impact on the interpretation of the different transport means. This is particularly obvious with respect to the environment and the higher number of those who say they are civil ecologists in Bern;
- The sedimentation of transport policies and urban planning policies over time has an impact on the appropriation of the different means of transport. The cities in which the public transport infrastructure has been destroyed and then rebuilt have populations that appropriate public transport less (a higher level of exclusive motorists). Similarly, in those cities where urban planning has been built around road networks the speed potential offered by public transport is less a part of the motivations of users (a lower presence of time-sensitive users).

Table 5.6   Weighting of the four types in the survey sample in each urban district

	France			Switzerland		
	Besançon	Grenoble	Toulouse	Bern	Geneva	Lausanne
Type 1 'Exclusive motorists'	34%	30%	36%	5%	21%	20%
Type 2 'Civic ecologists'	3%	3%	2%	14%	7%	5%
Type 3 'Constrained motorists'	30%	30%	36%	32%	34%	38%
Type 4 'Time-sensitive'	21%	27%	16%	40%	29%	26%
N	481	434	456	463	475	475

Conclusion

These four types illustrate the diversity of motility forms that underlie modal practice and their interaction with the destinations visited in everyday life. However, the search to minimise travel times is not always the primary motivation: on the one hand, speed potentials remain unexploited, and on the other, even when they are exploited, these potentials do not necessarily allow users to go faster. Finally, two conclusions come out of these analyses:

The use of a speed potential is far from automatic.   This is especially the case for public transport, which, even when it is faster than using the car, is not automatically used. This observation goes even farther: it has been shown that if car usage is restricted in any way, other places are frequented as a way of compensating for this restriction. However, this phenomenon remains confined to 'exclusive motorists' (type 1), who, although served by good-quality public transport, have a lifestyle based around the exclusive use of the private motorcar and a tendency to condition their destinations according to the possibility of using the car. Thus, not only are certain potentials not used, but certain actors have strategies that aim to avoid their use.

A speed potential is often appropriated to remain in a state of motility.   Thus, numerous respondents obtain access and skills via their motility – not to become mobile, but as a guarantee. This is a form of security that can be used in case something unpredictable happens in order not to be 'caught unaware'. The search for the broadest potential seems to be a very widespread practice as proven by the word 'practical' very often used to describe the car or public transport, whether or not either is used. This search is present among exclusive motorists who consider that public transport is practical and who have situated their homes near this network even though they have no intention of using it. This is also sometimes the reasoning behind civic ecologists who have automobiles and who use them little, but who also find it practical to have one 'in case' they are confronted with an unexpected situation.

These two conclusions show that, for the very specific example of modal choice, motility cannot be reduced to the mere quest to minimise travel times via speed. Among those interviewed, all of whom have access to both an automobile and public transport, many admitted to having obtained such access as a guarantee. This therefore provides me with the confirmation of the crucial importance of distinguishing speed potential from effective mobility: the speed potential offered by transport systems is appropriated as much to save time as to arm users against the risk of possible unforeseen events, and thus, does not necessarily serve to escape the bounds of territories.

Notes

1   In each urban centre, 500 people were interviewed using a random-quota technique. The cross-section is representative of the active population of the urban centre being studied (INSEE's definition in France and OFS's in Switzerland) according to sex, age and residential location (town-centre-suburbs). All the people interviewed have frequent-service (bus, tramway, metro) and/or quick (RER, metro-tramway) public transport less than 6 minutes away from their homes.

2   Two works have been published on this research: a scientific study (Kaufmann 2000) and a publication destined for urban transportation and planning professionals (Kaufmann and Guidey 1998).

3   For example, the number of journeys by public transport per person per year varies between 120 in Grenoble, 270 in Geneva and 470 in Bern.

4   As a definition of 'urban district' (agglomération urbaine), we have used that of the INSEE (for the French centres) and that of the OFS (for the Swiss ones) for the reference year 1990.

5   The number of journeys are rounded to the nearest ten. The calculations were carried out based on the 1993-94 timetable (the period the survey was carried out).

6   Table 5.5 shows, in particular, that when use of the car is quicker, more than 80% of respondents use it to go to work. On the other hand, when public transport is quicker, only 57% of respondents make use of it.

7   Traditionally, social perceptions are approached from the angle of positioning responses on bipolar scales. This method suppresses the possibility of the individual choosing the elements that seem to him or her the most relevant to describe the concept concerned. This appeared to me to be an inadequate way to study perceptions of different means of transport, effectively meaning that the designer of the questionnaire must choose bipolar positions in advance to submit to respondents. Given the limited level of knowledge in the area of social perceptions of different modes of transport, this option seemed arbitrary to me. I have therefore resorted to open questioning. This involved asking the respondent to give the three adjectives which best qualify public and private transport. They are, of course, later grouped.

8 On a methodological level, the construction of the typology is based on the results of a 'cluster analysis' and of a factorial analysis based on a corpus of adjectives aimed at describing the motorcar and public transport.

9 In the cases where the journey to a particular destination is unavoidable type 1 use their cars even when this use is rendered problematic by adverse traffic and parking conditions. A very significant proportion of them have a private parking space near to their place of work, which they pay for out of their own pocket if need be. If they do not have one they take the risk of parking illegally or pay a high price for moving from one parking space with a time limit to another during the course of the working day.

10 We have been able to observe that the constraint these users are subject to is stronger in Bern, Geneva and Lausanne than in the other urban centres studied. The result of this is an increased usage of public transport to get to the centre of these towns and cities.

11 The correspondence between these four types and the four ideal types for planning mobility presented in Chapter 4 is not obvious. The exclusive motorists were not included in the sample of interviews on planning as these interviews only dealt with train users. The civic ecologists are comparable to those seeking sensorial quality in their daily lives. The motorists forced to use public transport are split between the pre-planners and those open to opportunity, while those sensitive to time are related to the respondents seeking efficient alignment of activities.

6 What Inequalities?

Introduction

Do all aspirations to mobility find fertile ground for their realisation? Having explored the links between the intensity and the diversity of forms of mobility and the convergence between motility and mobility in Chapter 4, followed by the motivations underlying the transformation of motility into mobility in Chapter 5, I will now concentrate on inequalities in motility and mobility. The controversies raised in Chapter 2 concerning fluidification show how central this question is: to what extent are inequalities in mobility linked to the territory in question and to what extent are they a reflection of access to transport networks? Are these inequalities linked to projects that the actors have? Can they be linked to cultural models?

As with the other chapters that focused on research results, I have decided here to concentrate on a particularly revealing specific aspect rather than to broach the entire problem. Thus, I will use the example of the trend of urban growth in France as a basis for examining the inequalities that exist with respect to mobility. In particular, this will involve studying residential mobility and the use of the automobile. However, it does not mean focusing on the poorer sectors of the population and describing the difficulties these people encounter in acquiring their motility and transforming it into mobility, but instead implies highlighting the obstacles encountered by the entire population in its quest to carry out projects of mobility.

Urban growth makes for a particularly interesting topic when studied in this light, as a largely dichotomous vision of the city currently prevails in France in both academic and professional circles (Dubois-Taine and Chalas 1997). This vision functions as a social perception of urban reality, and opposes the historical form of the city (which should be conserved) and its emerging form (whose current development is said to be difficult to control).

- The former image, which may be referred to as the **historical city**, corresponds to centres that are centuries old and feature highly concentrated populations and activities. This form implies density and compactness, being designed as it was to accommodate city centres and neighbourhoods meant for access on foot; furthermore, it is noted for its economic, social and cultural diversity. Formed of contiguity and instant access, this urban model is the foundation of what is known as *urbanity*, a concept that is as much the product of place as of behaviour. The historical city is part of a spatial system marked by a strong distinction between the city and the countryside.

- The **outer suburban model** of the city is a product of urban spread. This type of city is built around large-scale transportation infrastructure, and is characterised by multiple central points reconstructed by its inhabitants. It represents the near-total spatial dispersion inherent in the urban lifestyles of the countries of the West. Its spatial limits escape urban morphology and depend instead on the flows of mobility; it is therefore no longer part of a city-countryside opposition, but rather the result of a gradual invasion of the land by urban development as and when improved accessibility allows.

The development of the outer suburbs that so marked the 1970s is often considered as the result of the population's aspirations with respect to lifestyle: this form of landscape is characterised by a near-absence of contiguity in activities, implies automobile use, and is noted for the individual character of its housing. It is common to read of the suburban city defined as the product of its suitability to the predominant values of the time, whether these are the wish to own property, the desire to live in a single-family home, or to travel by car for reasons of privacy or control over one's time and personal space. It is generally accepted that outer suburban development is a predominant and inevitable fact. Urban planning is therefore helpless in the face of this ground swell.

In this image of the city, the inner suburbs are reduced to a shadow. Inner suburbia is considered in this light either as an extension of the historical city, or as a transit zone between the two city forms; it certainly has no legitimacy in its own right. It is *de facto* considered a form of exile – a more or less demeaning context in which one finds oneself when one is deprived of access to both the historical city and the suburban habitat.

It is obvious that urban dynamics are a function of individual choice, but the residential mobility of households and their modal practice are also the result of choices made under duress. It is therefore a mistake to attribute people's behaviour wholly to their aspirations when it is in fact the product of opportunities and restrictions formed by the available options in general. It is not because forms of mobility are observed that they correspond to the motility of the population, which leads us to the question at the crux of this chapter: to what extent does the dichotomous vision of the city to which I just referred match the aspirations of the population, and to what extent does it reflect a system of constraints imposed on the actors?

The results of a survey in four cities

In order to examine this issue practically, I opted, along with my colleagues Christophe Jemelin and Jean-Marie Guidez, to compare a variety of urban configurations, an approach which provides an opportunity to contrast models that differ in their social make-up (socio-professional categories, lifestyles) and in their context (urban structure and available transport) and which furthermore are part of urban areas that vary in size. The areas chosen for the analysis were Paris-Île-de-France, Lyon, Strasbourg, and Aix-en-Provence. Districts with distinct types of social fabric were selected in each of the urban areas:

- Districts with the 'pre-automobile' urban model, designed with proximity mobility in mind, in each of the four cities: Les Gobelins (Paris), Charpennes (Lyon), Neudorf (Strasbourg), and Aix city-centre;
- Inner suburban models in Île-de-France, Lyon, and Strasbourg, i.e. the central districts of Ivry-sur-Seine, La Duchère and Hautepierre respectively;
- The *ville-nouvelle* (new city) model in Île-de-France: the centre of Evry;
- Outer-suburban models in Île-de-France, Lyon and Aix-en-Provence, i.e. Mennecy, Mions and Puyricard-Luynes.

A questionnaire administered by telephone was used to conduct the survey[1], which employed the quota method (sex, age, occupation of the household head, based on INSEE data from the 1990 population census)[2] in each district among a sample of 500 people aged 15 to 74 that included representatives from each type of neighbourhood studied. The results of the survey are given below.

The aspirations of the population are many and varied...

One of the main results of the survey is the confirmation that the current urban development in French cities whereby they are growing as a function of the automobile is not the product of generalized aspirations. For while it is true that there is a predominant model that exists and which links the car with social integration through connectivity and with single-family dwellings, other models were also identified:

- The reasons for choosing a particular type of transport vary greatly. This aspect, already identified in the previous chapter among a sample of respondents who had their own cars as well as public transport available near their homes, was confirmed here for the entire population. Not everyone wants to travel by car. While there was a marked preference for the car among the majority of those surveyed, there were also other inclinations which, while certainly not dominant, led some people to prefer to use public transport, to cycle, or to walk. These models were either based on respect for the environment, as in the case of the environmentally conscious, or on an attraction to the historical city and its public spaces, as in the case of those who naturally prefer alternative solutions. Thus, 19% of the surveyed population used the automobile but preferred either to use public transport, to walk, or to cycle, either because of their beliefs (6%) or for the qualities offered by these other means of transport (14%). A further 9% of respondents must be added as they never used a car by choice (they saw no need for this form of transport in their context).

• The spatial aspect of people's lifestyles differs greatly and cannot be summed up as a function of connectivity. There are many ways to appropriate a neighbourhood that have little to do with their morphology. What is more, spatial proximity does not appear to be a means of integration devoid of value, i.e. nothing more than compensation for the absence of access to motorised means of transport, nor even a vestige of the past in older central districts. Some respondents chose to appropriate the proximity of their neighbourhoods while developing a complex and diversified program of activities. It was as common for residents of new cities to frequent their own neighbourhoods as it was for those of central urban areas to do so; this was true even in outer suburban areas, where 30% of the population regularly or frequently attended activities in their home districts, a figure comparable to that observed in the inner suburban sites studied.

Table 6.1   Frequenting of resident neighbourhood according to type of urban model

	City centre	Inner suburbs	*Ville-nouvelle*	Outer suburbs
Frequently in neighbourhood	6%	5%	11%	5%
Regularly in neighbourhood	46%	26%	44%	25%
Occasionally in neighbourhood	42%	62%	40%	55%
Never in neighbourhood	6%	7%	5%	15%

• It is untrue that people wish to live only in the outer suburban areas of cities. Although some of the respondents who lived in the city centre did wish to live in the outer suburbs, the opposite was also true. It should be noted, however, that these desires concerning housing location were dichotomous: they either go in the direction of the city centre, or towards the outer suburbs. There was little desire to live in the inner suburbs, even among those who already did so. I therefore observed that the dichotomous vision of the city that opposes the historical district to its suburban form was not only present in professional circles and in urban research, but that it also exists among the population.

Table 6.2   Aspirations for residential location

Model	%
Would like to live in the city centre	46%
Would like to live in the inner suburbs	13%
Would like to live in the outer suburbs	41%

...but the urban dynamic sweeps them back to a dominant model

Although it is true that not everyone wants to either drive a car or live in the outer suburbs, the research has shown that there are constraints linked to the urban dynamic which push those who have other desires into adopting the dominant model. In particular, models that differ from the dominant model are alienated by restricted availability. The urban dynamic favours the car, and is therefore certainly the expression of individual desires compounded, but it also imposes itself as the only dynamic to the respondents who have other values.

Three aspects of the results illustrate this situation particularly well:

- First of all, among those who were predisposed to use several means of transport, respondents who preferred to use means other than the car because of their convictions, and respondents who preferred to use public transport rather than the automobile for the pleasure of using public spaces and being with other people were often forced to use the automobile, especially to go to work, because of mediocre public transport service to and from the workplace. This was also the case for travel to do shopping when no shopping centre was available close to home.
- I noted, with respect to how the respondents frequented their own neighbourhoods especially in the outer suburban areas, that the car was the means systematically used to appropriate the area around the home. In fact, these areas placed people in a situation of automobile dependence as defined by Dupuy (1999). A car thus became indispensable in these outer suburban areas studied in order to fulfill a somewhat complex schedule of activities, and in order to attend the facilities in one's own neighbourhood; the distances involved were too great to walk (especially with a geometric layout of streets that necessitates significant detours) and the available public transport did not allow for 'micro-mobility' (because this option existed during rush hours).

Table 6.3   Modal share of walking and cycling when travelling in one's own neighbourhood

City centre	Inner suburbs	*Ville-nouvelle*	Outer suburbs
72%	50%	19%	25%

- As far as the aspirations of the population for housing location were concerned, people were not wholly satisfied with the conditions of the housing market. Although our results highlight that the families of both white- and blue-collar workers aspired to live in the outer suburbs, the same results also identify an opposing viewpoint, i.e. those who wished to live in the centre of the city but who currently lived in the outer suburbs. This position was particularly true of three categories of the population: families who would have liked to own a large apartment and remain in a central city district but who were forced by the lack of housing suited to their tastes and their means to migrate and thus settle for the outer suburbs; households whose suburban location was the result of a compromise of diverging desires for residential location (for example in a couple where the man wished to live close to the city centre and but the woman preferred to live in a single-family dwelling); and teenagers and students who did not have access to a car and who had not chosen to live where they did, but yet who had to live with automobile dependence. It should be pointed out that the highest portion of respondents who wished to live elsewhere was among those living in the suburban context (and in the ville-nouvelle of Evry); this type of location seems to be less than desirable and, in some cases, was even stigmatised.

Table 6.4   Convergence between actual and desired residential location

	Lives in the city/inner suburbs		Lives in the outer suburbs	
	Would like to live in the city centre	Would like to live in the outer suburbs	Would like to live in the city centre	Would like to live in the outer suburbs
Owner	72%	28%	44%	56%
Prospective owner	63%	37%	45%	55%
Tenant	63%	37%	31%	69%

A dominant model

The results presented above show that French people have a dichotomous attitude about where they want to live, i.e. either in the city centre or in the outer suburbs. These results also show that there is a dominant model of aspiration in the population that links the use of the automobile with single-family dwellings. When integrated into an urban planning policy where this model is legitimised by policy procedures and incentives, it is sometimes difficult to choose another model if one has other projects. More precisely, four types of constraint can be identified.

The contextual limits of the possible realm

The first aspect is directly linked to the question asked at the beginning of this chapter, which was the meeting point between aspirations and context. It appears that certain aspirations do find fertile ground in the cities studied. The results suggest in particular that in these cities there is no housing available that would allow a family to live in a large apartment or a house located in the city centre at an affordable price, while there definitely is a demand for this type of housing. The results also show that there are few people who can do without a car every day, even though there is a fringe of the population that would like to be able to do so. This situation produces divergent motility and mobility: certain aspirations remain unrealised due to the lack of a context that would allow their realisation.

It needs to be pointed out that this constraint induced by the context is itself contextualised. In other European countries (or, in some cases, in other French cities if we follow the analyses of Marc Wiel (1999) on housing availability), cities are developed around other types of urbanisation that generate a different system of opportunities and constraints. Thus, families wishing to live for example in a house in the city close to the centre would have no trouble finding what they want in Great Britain, a country where the housing supply is essentially composed of terraced and semi-detached homes. On the other hand, families wishing to live in a detached home would have a more difficult time in the UK than in France, as this type of housing is very costly. The same logic can be followed concerning the use of the car; in Switzerland where the public transportation networks cover the surface of the cities both with respect to space and time, it is easier to do without a car in everyday life than in France.

Inequalities with respect to access

The second aspect concerns inequality of access. Some people find it difficult to realise their projects because their motility is limited by access constraints. This is a price-controlled mechanism; contrary to constraints related to context, availability is no problem in this case, but the supply is unaffordable. The most glaring instance I encountered in analysing the results was the desire to own a home among households of labourers and workers with modest wages; they are unable to afford such housing and usually live in multiple family buildings, whether low-income or not, on the outskirts of the city. This situation also appears among commuters who work in the city centre and use public transport to get to work, and who would prefer to use a car but cannot afford to pay for parking. Here again, price is an inhibiting factor. These situations also generate divergent motility and mobility as the desired mobility cannot be achieved because of motility restricted by access. This brings to mind the phenomena of territorial assignments that have been amply covered by research on suburban districts which shows that among the lowest-income households, there are many people who would like to live elsewhere. These results also concur with the research work on automobile dependence among poorer households (Froud et al 2000; Dupuy et al 2001) which demonstrates that a lack of access to an automobile drastically restricts potential everyday activities.

Compromises

The third type of constraint identified relates to compromises made within households. The mobility of some actors does not correspond entirely to their desires and aspirations following concessions made to their partner. This is yet another situation that results in divergent motility and mobility. This type of compromise was particularly identified in the area of choosing a residential location (in one couple, for example, the man prefers the city and the woman likes to be close to nature). This brings to bear the issue of power and of who is the decision-maker, of how to decide, and on the basis of which principles of fairness. Although the data collected does not provide a complete response to these questions, it does shed some light on the matter: more women than men wish to live in a context other than the one in which they currently live.

Although this type of situation was already identified in Chapter 4 with respect to life course, here I have quantified data that allows the population concerned to be precisely identified. The results show that women seem to be more subject to this type of constraint, but that they are not alone. Teenagers and young adults living with their parents are also sometimes quite dissatisfied with their residential location, especially when it is on the outskirts of the city. The dispersed fabric of the outer areas of the French cities studied imposes the intensive use of the automobile to develop planned activities, particularly in the evenings and at the weekends. Yet, teenagers and students often do not have access to a car, either because they do not yet have their driving licences or because they do not have their own cars and are thus forced to borrow one from their parents.

Appropriation dictated by culture

The fourth type of constraint is linked to the influence of culture on aspirations. This influence takes place through motility, and particularly through the aspect of appropriation. The attribution of values to the different contexts and means of transport guides behaviour and acts as a constraint. Thus, many families seeking to buy a large apartment will automatically exclude a certain number of suburban residential locations from their range of possible choices because they are seen as unattractive; this appropriation of context reduces their possible realm. More generally speaking, the dichotomous view of the city that currently exists in France in the scientific community, among professionals, among decision-makers and in the population constitutes a grid for interpreting the urban world that strongly shapes its appropriation. This view goes beyond the morphological aspects and equally concerns mobility. Thus it is that young adults are expected to appreciate the historical city for the amenities it offers, while families are supposed to want to live in the suburbs for the opportunity to have a garden and be close to nature. This is a veritable cultural model of the city and of its appropriation. This view is both hard to define and omnipresent, and affects supply, demand, public policy and the aspirations of the population.

This fourth type of constraint shows to what extent motility is not simply an improvised element that takes place among the actors, their interactions and positions, but also the expression of a culture. In this context, it should be noted that the cultural model described is not pan-European; the dichotomous view of the city that currently exists in France is in fact not present in most of the other European countries. The form of their cities is largely linked to the extent of their socio-spatial segregation, to urban development policies, and to existing laws. The examples of Great Britain, Germany and Switzerland are interesting in this sense in that they offer different rules and opportunities as far as aspirations and housing location are concerned. Each of these countries in its own way has developed urban development models that do without the dualist vision of the historical city in the hands of young people and singles and the outer suburban formula constructed for families.

The sedimentation of background policies

The four types of constraints developed above are linked, each in its own way, to the legislative body and to the incentives defined therein, as well as to public action and its sedimentation in the area of land development. Context, conditions of access, compromises within households and the appropriation by actors of the opportunities offered are fundamentally the reflection of the rules imposed by policy. The laws regulating housing, land development and city transportation constitute rules that define opportunities and are the expression of values. Similarly, the existing urban context (i.e. density, type of urban development, etc.), the available housing and transport, and the links between developed areas and transportation infrastructure can be considered as the sedimentation of past policies; this context defines a possible realm and contributes to moulding the city's image.

Underlying these different constraints are the tensions and cultural contradictions inherent in the policies that regulate the use of space. On the one hand, these policies are usually conceived as responses to the aspirations of the population, and particularly aim to satisfy the less fortunate members in a spirit of equal opportunity. On the other, however, they give legitimacy to a model of aspirations that prevents other projects from being carried out through a lack of favourable context. These observations show the tension that exists between values of equality and the respect of diversity and the promotion of the suburban model that is supposed to be so sought after. The commonplace confusion of actors' motility with their mobility (apparent in reasoning such as if 'people' go to live in the suburbs and use cars it is because this is what they aspire to do and therefore there is nothing one can do about it) makes these policies self-prophetic, providing a legitimacy that generates real urban dynamics.

Conclusion

The research presented highlights a culturally dominant model of city life, a model that is evident in the institutional and legislative systems as well as in the practices, values and aspirations of the actors. It integrates mobility in the sense that it proposes a path of mobility. That said, the results also show that this model is not shared by all, but that it is all-encompassing in that it is imposed on the entire population of the cities studied.

With respect to fluidification, I thus come to a paradoxical conclusion: fluidity exists only if one follows the culturally dominant model. On one hand fluidity appears to be significant among the actors who follow this model, since laws, incentives, context and dominant values all push for this model to occur. For those who do not follow this model, on the other hand, fluidity is limited; a set of constraints tends to push those with alternative projects in the direction of the dominant model.

In fact, fluidity resembles an illusion linked to the existence of this dominant model of city life. Fluidity appears in an ideological light that recalls the myth of a classless society that guarantees equal chances for all (see Chapter 1).

The example that was developed in this chapter shows that the context does not provide a favourable backdrop for the realisation of all the mobility aspirations of the population – far from it. It clearly appears that some of these aspirations are given legitimacy by a whole arsenal of situations and symbols, while other aspirations are disqualified.

Although one obviously has to be careful not to generalise on the basis of a specific example such as this one, the results hint at rules that are relatively restrictive in the end, and at a somewhat narrow scope of freedom among the actors. Considering that the housing supply and the transportation and urban development policies vary quite substantially from one European country and even one region to another, it may be assumed that this degree of freedom depends on the context.

Notes

1 The questionnaire was drafted by the IREC and the CERTU on the basis of the following existing and previously tested surveys (see Kaufmann and Guidez, 1998): the modal 'choice' surveys developed at the EPFL, the annual UNIVOX poll on housing (Switzerland) and the UTP-GART-CERTU surveys. New questions were added, for example, those concerning the use of the bicycle.
2 The questionnaire (about 20 minutes long) was distributed in November 1998 for the first phase (Île-de-France and Aix-en-Provence), and end-March and April 1999 for the second (Lyon and Strasbourg). The telephone interviews, carried out at the homes of the interviewees, took place from 6 p.m. to 9 p.m. during the week and from 10 a.m. to 7 p.m. on Saturdays.

7   The Production of Context

Introduction

The very idea of fluidification supposes that social and territorial structures take a back seat to a context that is capable of accommodating the most diverse aspirations. Chapter 6 showed, via a very specific example, that context was more than just a neutral backdrop for fluid forms of mobility. In the cities studied, one model of urban lifestyle takes precedence; it marks the territory through planning and the legislative apparatus, yet it corresponds to the aspirations of only a part of the population. This situation is an obstacle to certain forms of mobility which do not find a favourable terrain for their realisation. In this chapter, I will be examining again the issue of policies that affect the organisation of space by studying the production of context. How do social and political issues mark context? Via which process do we arrive at the development of contexts that facilitate certain aspirations and disqualify others?

The model of urban lifestyle that has become dominant in France is the expression of a particular culture. One does not have to travel far for this to become obvious: cities in France are conceived, inhabited and used differently from those in Germany or Great Britain. In the latter countries, cities are not conceived in terms of an opposition between the historical city centre comprising ancient buildings and the suburbs with their individual houses and scattered zones of activity; urban forms and their accessibility are built around other models. Thus, urban facilities, the conditions of access to these facilities and the value attributed to the different possible uses of the city are specific to each context (the context is not, by the way, necessarily defined along national border lines). With respect to motility, these observations reveal two facts: the first is that motility is formatted by the context, which pushes people to use certain forms of access rather than others and to acquire certain skills and appropriate the access forms in one way rather than another. Secondly, different aspirations and projects are likely to be realised with more or less ease depending on whether or not they fall into line with dominant lifestyles. These two findings illustrate the importance in social fluidity of the context and of its production. They also suggest that an analysis of the production of context would be best approached by means of comparisons.

Analysing public policy shows us that production of context is largely an issue of co-ordination between sectoral policies. For example, the impact on accessibility of a policy to develop public transport depends to a large extent on its co-ordination with urban planning policies: if urbanisation develops essentially outside the perimeters served by public transport, the impact will be much less than if urbanisation develops in areas where there is service. This fact is the result of the increasingly complex interdependence among sectoral policies, which require increasingly close co-ordination. The elements that affect the production of context are thus mainly found in the interfaces between sectoral policies.

In view of these considerations, I have chosen to base the analysis that is developed in this chapter on the results of a comparative study of the co-ordination of public policies affecting spatial organisation.

Chapter 7 is composed of three main sections. The first is devoted to a brief description of the survey data on which the analysis is based. The second presents the four aspects that together contribute to the production of context. The third section aims to highlight the incompatibilities between these four aspects.

The data

To approach this problem, I propose to use data from the Swiss participation in COST 332[1], a European research programme run jointly by different countries whose contributions are based on their national experiences. The focus of the programme is the co-ordination between public transportation policy and urban planning, and it seeks to discover the institutional and procedural responses to the question of how to integrate urbanisation with public transportation systems. This question, which has been the subject of a great deal of research over the last twenty years throughout Europe, has nevertheless remained largely unanswered. In general, we have the technical knowledge of what must be done in order to anchor urbanisation to train stations, but we do not know how to implement this politically. Structuring a city around public transport transfer points implies minute co-ordination between the regulating regional and urban transport policies and urbanisation policy, and this co-ordination is often difficult to implement. This applied example enables the production of context to be studied from three angles: values and political culture, institutions and legislative aspects and existing transport networks.

The research focuses on four cities in Switzerland: Basel, Bern, Geneva and Lausanne. There are three interesting reasons to compare these cities:

* First of all, they are located in different cultural areas: Basel and Bern are German-speaking, while Geneva and Lausanne are French-speaking. These cultures also have different perceptions of the importance of environmental issues; Basel and Bern are very environmentally aware, and

environmental conservation has much more political resonance in these two cities than in Geneva or Lausanne;

- Secondly, the Swiss model of federalism, which implies very strong decentralisation of power at the cantonal level, means that each of these cities has institutional and legislative specificities that produce local political cultures and define the modalities of political action. So in each of the four cities, the institutional rules are a little different. Geneva, for instance, has a tradition of public action in urban planning and there is comprehensive and restrictive legislation in this field that does not exist in Lausanne;
- Finally, the speed potentials offered by the different transportation networks are quite varied. The available regional rail transport is much more developed in Bern than in Basel or Lausanne, or especially Geneva. Moreover, urbanisation is planned around the train stations in the first two cities, while it has been structured around the roads in the latter two. The motorway networks are unequally developed also; while Bern and Lausanne are both located at important hubs, Basel and Geneva occupy a much less central position.

In comparing these four cities, I am looking at contexts that are likely to produce different systems of constraints and opportunities. To analyse the production of context concretely, case studies from each city have been retained. The objective of the case studies is to highlight the combination of rationales that guide action, and which are at work in the construction of a project involving both urban development and transportation dimensions. In these case studies, I focus as much on the constitution of the project, its context, and its objectives, as on the way in which the various actors involved in carrying out the project position themselves and interact with one another. In other words, it is an examination of the construction of the co-ordination process as a system of social action. Six case studies in all were selected:

Basel: the Claragraben Tram. This project, which has now been abandoned, consisted in building a tram line that was to connect with a transfer point. The objective was to provide additional transportation into the city centre from the north. The impact of the project was restricted to a very narrow area where population density was high.

Basel: the S-Bahn Green Line. This completed project consisted in constructing new, diametral rail transport using the existing infrastructure. It is part of a larger project to construct an S-Bahn network in the Basel region. The Green Line crosses the border into France, which complicated the decision-making process.

Bern: the Wankdorf Point. Currently underway, the project consists in building a central city hub comprising notably motorway access, an S-Bahn railway station, and a tramway terminus. This commuter complex is intended to house businesses, a new stadium, a shopping centre, leisure facilities, and a commuters' park-and-ride. The project location is presently sparsely populated; partnerships between the public and private sectors are involved.

Geneva: the Rhône Express Regional. This completed project consisted in renewing the light railway transport supply between the city centre and the western part of the canton of Geneva using existing infrastructure. The project, whose impact was regional, was located in a sparsely populated zone. The terminus of the line is situated on the French-Swiss border point of La Plaine. An extension of the line to the French city of Bellegarde is being studied, as is the intensification of urban development close to the railway stations.

Geneva: the Praille-Bachet-de-Pesay Point. This project, which is in the planning stages, consists in redefining the entrance into the city of Geneva from the South. Located at a motorway junction and in proximity to a public transport transfer point, the project includes plans to build a stadium, shopping centre, hotel and cultural facilities, commuter car park and a railway station. A key element is the heavy involvement of the private sector (the property developer of the stadium and shopping centre).

Lausanne: the extension of the Lausanne-Echalens-Bercher Line. The LEB extension project consists in extending a regional railway line to the city centre and developing an access point to public transportation at the new terminus. The project is situated in a densely populated area and aims to redistribute the urban concentration in Lausanne. It involves partnerships between the public and private sectors.

The methodology applied employed a semi-directive interview of those involved in each project. According to the size of the case study, the interviewees varied in number (from 4 to 15) and position (although they were mainly managerial level).

Four facets of context production

Analysis of the six case studies reveals four aspects that together produce a context: political goals, the means of intervention of government authorities, institutional architecture and the existing urban morphology. Several of these aspects work simultaneously to mobilise the motility of the actors.

The political objectives

The first aspect concerns environmental awareness and its impact on the definition of the goals of transportation and urbanisation policies and the co-ordination of these policies. In Basel and especially Bern, the objective of these policies is to limit the use of the automobile within the city for environmental reasons. The case studies show that this goal translates into very close co-ordination between transport policies and urbanisation policies: the means chosen to limit the use of the car is to anchor urbanisation around train stations. This is a policy that is therefore based on public transport accessibility. Its guiding principle is that different means of transport change the way an object is perceived. The automobile reigns in social values, and this is especially true in the case of investors. It is also a fact that road networks are so tightly woven that the lack of co-ordination between urban and regional transport and town-planning policies is of little consequence, given that housing developments have 'naturally' grown up around the road networks. On the other hand, linking urbanisation to public transport would imply a voluntarist policy, since infrastructure and operating costs are high and service is by definition intermittent. This policy is based on the principle that actors' motility is out of step with the political will to reduce the use of the car. In order to get around this opposition, public action uses the aspect of access: by making it difficult to access certain places by car, the use of other means of transport is promoted.

In Geneva and Lausanne, where awareness of environmental issues is much less marked, transport policy does not have as its primary objective to limit the use of the automobile, but rather to ensure the complementary use of different means of transport, based on the idea of modal choice. Transport policy is thus based in some cases on automobile accessibility, especially with respect to the entry points into cities, and in other cases on multimodal accessibility, that is to say, with good public transport service and the possibility of access by car. Conservation of the environment is not enough of a concern for the authorities to envisage imposing restrictions on users. The transportation and urbanisation policies in Geneva and Lausanne take up the motility of actors by using appropriation. The basic idea is to consider that if there is a competitive alternative to the automobile, people will take it into account in their choices and that in this way city automobile traffic will be reduced. These results clearly show the impact that political objectives can have on producing a context.

At first sight, the Geneva/Lausanne option is more fluid, while its counterpart in Basel and Bern seems more restrictive. In the first case, the actors are placed in a situation that forces them to adopt certain practices; in the second case, alternative options are proposed, and each person is free to choose what suits him or her best. And yet, on closer inspection, things are not quite as cut and dried. Chapter 5 showed how the use of speed potential is far from being the result of automatic reactions, especially with respect to public transport. It can then be deduced from this observation that the co-ordination between urban development and transportation as it exists in Geneva and Lausanne contributes to producing a context that is likely to impose the use of the car on actors who do not wish it, by means of the mechanisms that were highlighted for the cities of Paris, Lyon, Strasbourg and Aix-en-Provence (see Chapter 6). Ultimately, the political choices made in Basel and Bern and those made in Geneva and Lausanne produce sets of contextual constraints that differ from one another, but that are difficult to classify in terms of degree of fluidity.

The mode intervention by authorities

The second aspect concerns the means that public authorities use to inter-vene in the implementation of co-ordinated policies in transportation and urbanisation. Three ways of perceiving the authorities' role were revealed: the planning state, the incentivising state, and the 'offering' state.

* The first category, *the planning state*, consists in defining a field of possi-ble action for urbanisation strictly limited by plans for zones according to their accessibility. In the case studies, the Wankdorf project in Bern uses this means of intervention. The Swiss part of the Regio-S-Bahn Green Line was also conceived in this perspective.
* The second type, *the incentivising state*, consists in facilitating the urbani-sation of certain places because of their accessibility using notably finan-cial incentives. The La Praille transfer point in Geneva is the result of an *incentive planning* scheme wherein the state facilitates settlement in to this area by offering better financial conditions than in an industrial area.
* The third approach, *the 'offering' state*, consists in deploying the transport infrastructure so as to provide solutions of accessibility that are intended to structure urbanisation. Planning by offer largely underpinned the proj-ects to extend the Lausanne LEB and the RER line from La Plaine to the city of Geneva. In these two projects, the authorities confined themselves to developing public transport availability by postulating its attractiveness to investors.

In all these cases, the way in which the state's role is conceived refers to a system of values that can be related to political divisions. The planning state model resembles the republican model of government where decisions taken do not necessarily take into account the economic milieux or the considera-tions of other lobby groups. In Switzerland, it is often this model used to integrate transportation with urbanisation that is championed by environmentalists. The incentivising state is in tune with those models of govern-ment in which the state negotiates with the actors and adapts the founding principles to influence the decisions of private investors. In Switzerland, case studies suggest that this method has the political backing of parties that place themselves at the centre-left of the political spectrum. The offering state, for its part, implies that an investment policy in infrastructure is likely to influence the structure of urbanisation in a decisive manner. In the cases studied, this model is upheld by the political right.

Each of these ways of planning produces a context with specific opportu-nities that appeal to actors' motility in different ways: the planning state channels motility in the name of common interest by designating a territori-ally defined context for it. Motility in this case is regulated by political objectives that may even constrain it. The incentivising state pushes actors to adopt an economically rational form of conduct. In so doing, it requires the actors to have fairly developed evaluation skills and incites the population to appropriate the context according to the rational choice model. The offering state counts on the use of the speed potentials offered. It gambles on the structuring effect of infrastructure and thus attributes an important role to actors' motility, since these different facets will determine the use of what is offered. In this last option, the policy implemented is regulated by the actors' motility: the success or failure of the political choices made depend on the motility of the population as well as on investors.

Institutional architecture

The third aspect is institutional architecture and its formalisation by legisla-tion. The case studies have shown that this aspect defines the field of possi-ble action for co-ordinated policies and, at the same time, for producing the context. Political objectives and the means of intervention of the authorities are implemented through the institutional architecture and the body of legis-lation. This operation is not at all taken lightly and can mark the content of a project, and facilitate, hinder or even prevent a co-ordinated policy. Four factors linked to institutional architecture appeared in the case studies as being relevant.

- The 'dispatching' of services in an administration can reinforce certain links and weaken others; in this way, co-ordination and coherence in pub-lic action can be facilitated or hindered. When administrations in charge of transport and land development are part of the same department, co-ordination is easier (as is the case in Bern); when on the contrary these administrations belong to different departments, co-ordination is more complex, and this is exacerbated when the judges heading the departments are from different ends of the political spectrum (as in Geneva).

- Likewise, the vertical distribution of skills can either facilitate or hamper co-ordination between urban planning and transportation policies. The case studies show in particular that complete control of distribution plans at the cantonal level (macro-local scale) facilitates this co-ordination, while control at the communal level (micro-local scale) acts against it, because the number of actors involved is multiplied and co-ordination becomes piecemeal (the case of Lausanne).
- The transparency of the dispatching process among public actors and the degree of reciprocal impermeability of the political and technological spheres also have an impact on co-operation among these actors. An unambiguous organisational chart, combined with the separation of political and technological stakes, facilitates cooperation. This is the major asset that was implemented in Bern for the Wankdorf project. The institutional vagueness that appears throughout the case studies is an enemy of co-ordination: duplications and poorly defined control of the project are sources of conflict that encourage institutional stand-offs to the detriment of a project-centred rationale.
- Financing opportunities weigh heavily on the strategic options retained in an infrastructure or urban development project with respect to the co-ordination between urbanisation and available transport. The case studies show that financing opportunities are an obstacle to the integration of urban planning with public transport infrastructure because they are divided according to sector. There is law that comprises regulations allowing projects for co-ordinating infrastructure with the public transport sector to be financed. Of the cases studied, the extension of the Lausanne-Echallens-Bercher line shows this point most clearly: the federal financing opportunities for this project are subject to the Railway law, the 'RAIL 2000' law and the edict on the separation of traffic, which implies that this project must belong only to the area of transportation in order to benefit from subsidies.

These four factors indirectly mobilise the motility of actors. By defining the limits of the fields of possible public action, they more or less allow for a context likely to accommodate different aspirations and lifestyles. In this respect, they will more or less provide for the expression of diversified lifestyles.

The pre-existing urban morphology

The final aspect that I have identified as having an effect on the production of context has to do with the sedimentation of policies that have spatial consequences. The case studies have shown that the political objectives and modes of intervention of public authorities depend on the pre-existing context. The pre-existing territorial policies thus induce the structuring elements of the future spatial policies – both in practical terms and in terms of representation – which means that it is sometimes difficult to go against the grain of established spatial dynamics.

A twofold movement has been observed: on one hand, the morphology of the land shapes the choices for co-ordination and the motility of actors, and on the other hand, urban planning decisions taken previously create this morphology. The configuration of a living area is not only a determining factor in the co-ordination between transport and urban planning in that it defines the way in which the dimension of space is perceived by the actors, but it is also the result of such a process. Current physical context is partly the product of past policies concerning the integration of development and transport. The size and shape of light railway networks are especially influential, since they represent the spatial outline of a co-ordinated policy.

More specifically, the case studies have shown that the existence or lack of rail infrastructure in a city influences not only decisions relating to public transport policy, but also urban planning decisions. Integrating urban planning with the planning of railway infrastructure is both more ambitious and more complicated than implementing the same policy in a city where the railway infrastructure already exists. The case of the Canton of Geneva illustrates this well: although Geneva has never been an important hub for railway transport, it did have an impressive tramway network which was almost totally dismantled during the 1950s and 1960s. Therefore, in the minds of certain planners, urban development in this city can no longer be integrated with heavy public transport infrastructure, which would have to be rebuilt. Indeed, as testified by master plans for urban development in Geneva since the 1970s, planning authorities find it hard to imagine integrating urban development with anything other than road networks. The option of structuring the city around a railway appears unimaginable, as if the lack of a real railway network stands in the way of creating one.

Incompatible elements producing context

The body of case studies shows that the objectives of co-ordination, the principles of action retained to attain these objectives, the institutional architecture and the pre-existing state of transport and urbanisation infrastructure often lead to unwanted effects that produce a context that does not correspond to the objectives set. This situation is the result of an incompatibility among the four aspects highlighted above. The incompatible elements revealed relate to socio-political and socio-cultural contradictions.

The contradictions between institutional architecture and the objectives and principles of action

In several of the case studies, it appeared that the institutional architecture was not in line with the political objectives or the principles of action retained to attain these objectives. This was the case for the extension of the LEB railway line in Lausanne and for the La Praille-Bachet transfer point in Geneva. In both cases, the objectives of co-ordination were badly handled by the financing of the project and the sectoral division of the work to be done. The extension of the LEB, categorised as a railway project, obtained financing as infrastructure, and the project was headed by the transportation department of the Canton of Vaud. Similarly, the construction of the La Praille-Bachet transfer point obtained subsidies as equipment (for the stadium) and was headed by the urban planning department. In both cases, public financing, divided on a sectoral basis, contained no 'bonus' for co-ordination; on the contrary, co-ordination increased project costs and was thus banished. This abandonment of attempts at co-ordination was facilitated by the fact that the leadership of the projects was assured by a department that did not deal with the other area of activity, i.e. transport or urban planning.

The political objectives of co-ordination are unable to be attained because the financing opportunities are unsuited to them and the dispatching of services in the varidus departments is an obstacle to them. The result is that a context facilitating the use of the automobile is produced by default. The automobile, because its specific characteristics of mobility (individual rapid means of transport), is indeed the only means of transport that can compensate for a lack of co-ordination.

The contradictions between political objectives and principles of action, and actors' motility

The different modes of intervention used by authorities, whether in the form of planning, incentives or 'offering', do not all allow for the objectives set for co-ordination between urban development and transport availability to be reached.

The case studies highlighted in particular that the objective of integrating urbanisation with the public transport infrastructure is incompatible with the 'offering' mode of intervention by the state. Even if such a combination is theoretically imaginable (and is in fact applied in cases where the pre-existing urban morphology strongly encourages the use of the automobile), its realisation is compromised by the 'structuring' effect of the road infrastructure on urbanisation. The case studies involving the La Plaine RER light railway in Geneva, the LEB in Lausanne and the S-Bahn Green Line in Basel are very revealing of this incompatibility: in all three cases, the realisation of the new form of transport had only a minimal impact on urban development, which continued to build itself around rapid road networks. Households and businesses give preference to road accessibility when choosing a location. Their motility pushes them for the most part towards the automobile because of the club effect linked to the fact that the pre-existing urban morphology encourages the use of the car. The result is the production of a context that favours automobile dependence, which is the opposite of the political objectives initially set.

This result shows how crucial it is to take into account values when considering context production: to affirm that urbanisation must be centred around train stations for reasons of environmental conservation implies overcoming cultural contradictions and requires modes of intervention likely to channel the motility of some people by using rules that encourage or impose location in places with good public transport service.

Governance in question

The examination of incompatible elements illustrates that although policy is central to the process of producing a context, the type of governance applied is just as important. The authorities can perfectly well develop a coordinated project that aims to integrate urbanisation with infrastructure in partnership with different private actors, but based on the implementation of means that revent the objective of integration from being achieved because the motility of the actors was not taken into consideration. In other words, there can be social and political consensus with respect to technological solutions that are inappropriate to attain the objectives set, because they do not match the users' logic of action. Having the ambition to structure urbanisation around public transport infrastructure – as is the case in Switzerland – necessarily implies applying a policy aimed at channelling the motility of the actors, such as that implemented in Bern, in order to transcend the contradictions between socio-political values and socio-cultural values.

Conclusion

The above analysis of the production of context based on the specific example of co-ordination between urban development and transport in four Swiss cities leads to a central conclusion: *the production of a context that is restrictive for the actors involved, in the long run, is the fruit of policies that are regulated by the motility of the actors alone.*

Given the existing value systems, users who have a choice will mainly choose to use the car. Road systems and heavy public transport do not hold the same interest for investors – so different are they in this respect, in fact, that they cannot be considered in the same way. Although new cities are built around bypasses and motorway intersections without having to plan around anything other than existing availability, developing urbanisation centred around public transport infrastructure on the contrary implies having recourse to planning or incentives that 'make the rules' in such a strict manner that it is impossible to circumvent them.

This conclusion may seem paradoxical. A policy based on the creation of infrastructure alone may limit the realm of possibility in terms of mobility. In this case, since investors have a tendency to be interested only in road infrastructure, all the other available means are gradually disqualified since the city is built around road access. On the other hand, the case study of Bern shows that a policy of restrictive access with respect to the car, founded on the stringent integration of the developed areas with public transport availability, is able to encourage the development of a large range of aspirations. The aspects highlighted by the case studies as being central to the production of context and the incompatibilities of these aspects suggest that the mode of intervention of the authorities is used as a strategic resource to reconcile the contradictions between socio-political and socio-cultural aspects. The inappropriateness of the means employed and the political objectives announced is understandable and enables both to justify a policy aiming to reduce the use of the car that is supposed to be voluntarist, as it is based on significant amounts of investment, and to preserve the interests of the private investors who wish to be located only where there is road access.

Finally, throughout this chapter fluidity appeared as an ideology at the service of the dominant actors. Developing alternative transport supply while saying 'it is up to the investors and users to choose' is tantamount to promoting the use of the car, considering its dominant position in daily mobility and the club effect that this produces.

Note

1   This research was carried out by a team comprised of Yves Ferrari, Dominique Joye, Vincent Kaufmann and Fritz Sager. A report on it was published at the IREC in 2001 under the heading 'The Co-ordination between Transportation Projects and Land Planning: in Basel, Bern, Geneva and Lausanne'.

8 Conclusion:
Towards a Network Solidity?

To what extent can the speed potentials procured by technological systems of transport and telecommunications be considered vectors of social change? Having discussed the relevant writings, proposed a conceptual tool to define mobility and developed empirical analyses using the concept set forth, it is time to conclude.

Given the prospective nature of the entire approach, this conclusion will take the form of an opening. After all, it would not be reasonable to take a clear position in arguments that largely escape the field of the analyses realised – even less so since the approach I am proposing opens new doors for areas of research in social science.

This body of work is devoted to proposing the notion of motility and to testing its heuristic virtues. Taking as the point of departure the position supported in a great many writings on the fluidification of societies under the impulse of speed potentials procured by technological transportation and telecommunications systems, I have shown the need to develop the link between theoretical work and empirical research in order to be able to speak of the impact on territory and society of speed potentials. I have pointed out that the lack of conceptual tools to distinguish available speed potentials from spatial mobility proper explains the dearth of dialogue between theory and empirical work, as well as the often normative and blanket nature of the positions adopted by researchers with respect to social fluidification.

In order to bypass this situation I proposed the concept of motility. Motility can be defined as the operation of transforming speed potentials into mobility potentials. It is therefore the way in which an actor appropriates the domain of what is possible in the area of mobility and makes use of it to develop his or her projects. By differentiating explicitly mobility potential from mobility, strictly speaking, the notion of motility allows me to distinguish social fluidity, from spatial mobility, and spatial mobility from the motivation for action which underpins them. In so doing, it poses in a new way the problem of the possible social fluidification produced by speed potential procured by transport systems, and the compression of time and space that accompanies it, to be studied through four questions:

- Are people more 'free' when they are more mobile? Or, *is the convergence between motility and mobility stronger for those who are more mobile?*

- Do actors seek maximum speed in their mobility? Or in other words, *to what extent does motility converge with the search for maximum efficiency in one's mobility?*
- Do all aspirations of mobility find favourable terrain for their realisation? In other words, *is it more difficult for certain types of motility to be transformed into mobility?*
- Do speed potentials generate social fluidity? Or, *to what extent do speed potentials diminish or increase social inequality?*

I propose to develop this conclusion in three points, in response to these four questions, by highlighting the contributions of the conceptual tool proposed and the research paths opened by the results of my work.

Social structures become networked without becoming more fluid

An initial conclusion arises from the series of analyses developed from the concept of motility: we are witnessing the birth of a new form of social and spatial structure based on the use of the speed potentials procured by technological transportation and telecommunications systems, but this new structure is no more fluid than that which it is replacing. This first conclusion came about as a result of two observations:

- *Actors are not necessarily more mobile because they travel faster and farther.* The analyses showed that those who use the speed potentials procured by technological systems are in general no more mobile than those who do not, whether this is in terms of the number of journeys or in terms of the diversity and complexity of their mobility. On the contrary, the interviews presented in Chapter 4 showed that among the respondents with the most diversified and complicated mobility forms there were many people who make only fairly selective use of the speed potentials made available by transport systems. Forms of mobility that are connex, reversible and ubiquitous and which use the speed potentials procured by transportation technology do not by definition mean increased mobility;
- *Actors who are 'free' do not necessarily travel fast and far.* The analyses developed have shown that technological transport and telecommunications systems do not free people from social constraints. Many of those who make the most use of speed potentials lead daily lives limited by numerous constraints. The results of the interviews presented in Chapter 4 indicated that those who make the most use of speed potentials often place their work at the heart of their existence. Their highly developed mobility is often a more or less direct response to the flexibility that companies expect of their managers. Connex, reversible and ubiquitous forms of mobility seem more like submitting to structures than 'getting

away'. Such mobility forms are often the result of compromises between career and family. In addition, the empirical research showed that those people who have more flexibility in their life courses generally combine mobility with sedentarity. There are many possible combinations, but they all have one thing in common: a fixed base point be the residential location, the neighbourhood as a uniting venue of everyday life, or the absence of commuting.

In sum, it would appear that the phenomena of connexity, reversibility and ubiquity, often considered in current literature as signs that our societies are becoming more fluid, are instead a mandatory step towards attaining social insertion. These forms of mobility are indeed increasingly necessary to be able to juggle the different aspects of social life; made possible by technology, they have truly freed people from certain constraints of daily life, but have created other restrictions. By allowing actors to combine and reconcile what was once irreconcilable, these mobility forms have widened the realm of the possible for actors, but at the same time have made them dependent. This aspect has been entirely ignored by many advocates of the network model – and even more so by supporters of the liquid model – whose vision of connexity, reversibility and ubiquity is limited to the expression of a new 'freedom'.

This initial conclusion illustrates the need to distinguish between spatial mobility and social fluidity. They belong to two different spheres of reality, and the first is definitely not a good indicator of the second. The confusion between spatial mobility and social fluidity is certainly linked to the prestige associated with the notion of speed. On several occasions, the empirical data I used allowed me to observe that the most connex, reversible and ubiquitous forms of mobility are associated with increased mobility and 'freedom', as opposed to forms of mobility that are contiguous, irreversible and unifying. The symbol associated with the automobile is especially revealing in this respect: its core associates speed and freedom in space and time.

This result constitutes a contribution to the debate on social fluidification. It suggests that *the networking of society is indeed not accompanied by its fluidification.* Although this affirmation naturally must be validated by other analyses, it nevertheless opens at least one new path for investigation – that of the birth of new factors of social differentiation that are no longer built around defined territorial limits, but instead around space and time.

Mobility as a value that reveals cultural contradictions

The above considerations lead me to the second point of my conclusion. Mobility, far from being a neutral practice, takes on the characteristics of a value. Yet, the debates on social fluidification have shown that many analysts confuse the social phenomena with which they are faced in their analyses with the social representation of these phenomena. All too often, the normative dimension of mobility takes over from its conceptual aspect – that which allows practices to be observed. This is the case especially each time that significant meaning is attributed first and foremost to the different forms of mobility.

Mobility is a value with its own inherent differentiations. Juggling with mobility can allow someone to acquire a social status. Connexity, reversibility and ubiquity are socially prized. All three are related to power through speed and by the ability to act from a distance; they give the illusion of escaping from social and territorial structures. Contiguous, irreversible and unifying forms of mobility, on the other hand, are socially devalued, especially because they do not necessarily require technological means. The research presented highlighted that these differentiations between forms of mobility reveal cultural contradictions that are likely to cause tension between contradictory motivations, inequality with respect to realising aspirations, or to strategies aimed at reconciling the contradictions:

- *Tension between conflicting motivations.* I encountered this case in the analyses presented in Chapter 5 with respect to the logics of action underlying modal practice. Even when public transport is faster than the automobile, public transport is not necessarily used. To take the same logic further, actors develop strategies to avoid using public transport, particularly by giving preference to destinations that can be accessed by road. The attribution of the qualities of comfort and independence to the automobile are at the root of this phenomenon. In terms of values, there is thus a conflict between the minimisation of travel times (speed) and the 'freedom' gained by using a car.
- *Inequalities with respect to realising aspirations.* The research results presented in Chapter 6 show that the value attributed to certain forms of mobility is supported by public action and the legislative body of texts. This means promotes a certain model (individual housing), which ends up becoming the dominant model imposed on the entire population. As long as one subscribes to the dominant model, it is fairly easy to realise one's desires as to residence. As soon as one steps out of this model, aspirations are limited by a set of constraints and incentives which pushes actors to adopt the dominant model which nevertheless contradicts the stated political priority with respect to durable mobility in the contexts studied.
- *Strategies to reconcile cultural contradictions.* The analyses in Chapter 7 show that a policy that aims to limit the use of the car in the city is only efficient if it includes restrictive measures with respect to the automobile. Given this situation, the authorities sometimes develop measures which

they know are inappropriate for reducing the use of the car, such as investing heavily in public transport, but which allow them to justify a voluntarist policy (since it is based on significant amounts of investment), while preserving the idea of 'free choice' as to the means of transport in the city.

Faced with these results, mobility appears in an ideological light. This fundamental aspect discredits the rhizomatic model that takes for granted the equation speed = mobility = fluidity, as well as in general all the interpretations of mobility as a vector of liberation from social structures. *Mobility is polysemic and does not itself reveal what underlies it.*

This second conclusion illustrates the need for sociological approaches that focus on the combination of motivations for actions. In so doing, the conclusion shows the pertinence of the concept of motility, which deals precisely with understanding how and why speed potentials are transformed, or not transformed, into mobility. This opens up an entire field of study, so far little explored: that of the sociology of latency or of potential practice.

Motility is a source of wealth

The need to open the 'black box' of mobility leads me to the third aspect I would like to develop in this conclusion. Spatial mobility is not an interstice, or a neutral liaison time between a point of origin and a destination. It is a structuring dimension of social life and of social integration. There are increasing numbers of ways to be mobile, and this multiplication is the result of the speed potentials made possible by technological systems. Throughout the chapters, it became clear that mobility is a resource to which actors have recourse to appropriate the many options open to them within the given context.

Just as money is related to financial capital, to knowledge and to its transmission to cultural wealth (culture in the sense of education and not in the anthropological sense), so relationship networks relate to social capital and mobility relates to motility and constitutes a capital. Someone may be rich or poor in this form of capital, and can especially possess this capital in different ways. *Unlike the cultural, financial and social capital forms, which deal mainly with hierarchical position, motility relates as much to the vertical as to the horizontal dimensions of social position.* Not only does the motility capital identify a new form of social inequality, it allows lifestyles to be distinguished with respect to space and time. Motility appears in particular as an indispensable resource to enable people to get around the many different spatial constraints that bind them. Quality of life often depends on the ingeniousness of the solutions invented and applied.

These considerations, and especially the gradual widening of the realm of choice in the field of mobility, suggest that motility is a form of capital whose importance is on the rise:

- On one hand, the multiplication of possible options introduces differentiation where there once was none. Not only does being mobile often imply having to choose between alternatives, but the realm of choice and the skills needed to make this choice are also constantly changing. Actors increasingly find themselves facing access choices (which they should adopt or not), choices of skills (to be acquired or not) and of appropriation (of the analysis of the interest of one means of communication over another) whenever they wish to be mobile;
- On the other hand, the multiplication of speed potentials and their evolution is feeding the production of novelty via forms of mobility that are prized and those that are despised. By their impact on practices, these new rules are likely to produce social change and thus reinforce the importance of motility as a capital.

The empirical data analysed in Chapters 4 and 6 shows that in general, actors seek to arm themselves with the broadest possible mobility by acquiring skills for and access to the largest number of technological systems. The data shows in particular that motility is not necessarily there to be transformed into mobility; many actors acquire access and skills, not to be mobile, but rather to insure themselves against all kinds of risks, to be sure not to be caught short in very different situations which vary from daily life to their professional careers.

Fundamentally, motility shows us that if conventional forms of social science analysis are less able to differentiate than in the past, it is not because the social and territorial structures are being replaced by a more fluid world that is either more or less stratified or binary, but rather because these categories of analysis are no longer able to read a changed social reality.

What is certain is that motility invites us to refresh our view of the world and the tools we use to analyse it.

Bibliography

Albertsen N. and Diken B. (1999) *What is 'the social'?* Working Paper 145, Roskilde University: Department of Geography and International Development Studies.

Ascher F. (1995) *Métapolis ou l'avenir des villes*, Paris: Odile Jacob.

Ascher F. (1998) *La République contre la ville*, La Tour d'Aigues: éditions de l'Aube.

Ascher F. (2000) *Ces événements nous dépassent, faignons d'en être les organisateurs*, La Tour d'Aigues: éditions de l'Aube.

Ascher F. (2000) Postface: les mobilités et les temporalités, condensateurs des mutations urbaines, in: Bonnet M. and Desjeux D. (eds.) *Les territoires de la mobilité*, Paris: PUF, 201-214.

Augé M. (1992) *Non-lieux*, Paris: éditions du Seuil.

Badie B. (1995) *La fin des territoires*, Paris: Fayard.

Bailly J.-P. and Heurgon E. (2001) *Nouveaux rythmes urbains: quels transports?* La Tour d'Aigues: l'Aube édition.

Balandier G. (2001) *Le grand système*, Paris: Fayard.

Bassand M. (1980) *Mobilité spatiale*, St.-Saphorin: Georgi.

Bassand M. (2000) 'Métropoles et métropolisation', in: Bassand M. Kaufmann V. and Joye D. (eds.) *Enjeux de la sociologie urbaine*, Lausanne: Presses polytechniques et universitaires romandes, 3-16.

Bauman Z. (2000) *Liquid modernity*, Cambridge: Polity Press.

Beaucire, F. (1996) *Les transports publics et la ville*, Toulouse: Editions Milan.

Beck U. (1992) *Risk Society*, London: Sage.

Beck U., Giddens A., and Lash S. (1994) *Reflexive Modernization – Politics, Tradition and Aesthetics in the Modern Social Order*, Cambridge: Polity Press.

Bell D. (1978) *Les contradictions culturelles du capitalisme*, Paris: PUF.

Bellanger F. et Marzloff B. (1996) (eds.) *Transit*, éditions de l'Aube: La Tour d'Aigues.

Boden D. and Molotch H. (1994) 'The compulsion of proximity', in: Friedland R. and Boden D. (eds.) *Now Here – Space, time and modernity*, Berkeley, Los Angeles: University of California Press.

Bonnet M. et Desjeux D. (2000) *Les territoires de la mobilité*, Paris: PUF.

Bonvalet C. and Brun (1998) 'Logement, mobilités et trajectoires résidentielles', in: Segaud M., Bonvalet C.X. and Brun J. (eds.) *Logement et habitat – l'état des savoirs*, Paris: éditions la découverte.

Bordreuil J. S. (1997) 'Insociable mobilité?', in: Obadia A. (ed) *Entreprendre la ville – Nouvelles temporalités – nouveaux services*, Colloque de Cerisy, La Tour d'Aigues: l'Aube éditions, 215-227.

Boudon R. (1995) *Le juste et le vrai*, Fayard: Paris.

Brulhardt M.-C. and Bassand M. (1981) 'La mobilité spatiale en tant que système', in: *Revue suisse d'économie politique et de statistique*, Vol. 3/81, 505-519.

Castells M. (1996) *The rise of the network society – the information age*, Blackwell: Oxford.

CEMT (1996) *Réduire ou repenser la mobilité urbaine quotidienne?*, Table ronde 102 de la conférence européenne des ministres des transports, Paris.

Chalas Y. (1997) 'Les figures de la ville émergente', in: Dubois-Taine G. and Chalas Y. (eds.) *La ville émergente*, La Tour d'Aigues: éditions de l'Aube.

Champion, A. (ed.) (1989) *Counterurbanization*, London: Edward Arnold.

Choay F. (1994) 'Le règne de l'urbain et la mort de la ville', in: *La ville – art et architecture en Europe 1870-1993*, Paris: Centre George Pompidou, 26-35.

Claisse G. and Duchier D. (1993) *Des observations d'effets TGV: réflexions méthodolo-giques*, 6ᵉ entretiens du centre Jacques-Cartier sur Villes et TGV, Lyon, 5-11 décembre.

Couclelis H. (1996) 'The death of distance', in: *Environment and Planning B – Planning and Design*, Vol. 23, 387-389.

Cuin Ch. H (1983) *Les sociologues et la mobilité sociale*, Paris: PUF.

Cwerner S. B. (1999) *The times of migration, a study of the temporalities of the immigrant experience*, PhD in Sociology, Lancaster University.

De Boer E. (1986) *Transport Sociology – Social Aspects of Transport Planning*, London: Pergamon.

De Certeau M. (1980) *L'invention du quotidien – tome 1, Arts de faire*, Paris, Union générale d'éditions, collection 10-18.

De Haan A. (1999) 'Livelihoods and Poverty: the role of migration – a critical review of the migration literature', in: *The Journal of Development Studies*, Vol. 36 N° 2, December 1999, 1-47.

De Singly F. (2000) *Libres ensemble – l'individualisme dans la vie commune*, Paris: Nathan.

Degenne A. and Forsé M. (1994) *Les réseaux sociaux*, Paris: Armand Colin.

Deleuze G. and Guattari F. (1987) *A thousand plateaus*, Minneapolis and London, University of Minnesota Press.

Deleuze G. and Guattari F. (1994) *Difference and repetition*, London: The Athlone Press.

Deleuze G. and Parnet C. (1996) *Dialogues*, Paris: Flammarion.

Diken B. (1998) *Strangers, Ambivalence and Social Theory*, Aldershot: Ashgate.

Du Bois P. (ed.) (1983) *Union et division des Suisses – les relations entre Alémaniques, Romands et Tessinois aux XIXᵉ et XXᵉ siècles*, Lausanne: L'Aire.

Dubet F. (1994) *Sociologie de l'expérience*, Paris: éditions du Seuil.

Dubet F. (2000) *Inégalités sociales et Société-monde*, Congrès de l'Association internationale des sociologues de langue française, Québec.

Dubet F. et Martucelli D. (1998) *Dans quelle société vivons-nous?*, Paris: éditions du Seuil.

Dubois-Taine G. et Chalas Y. (1997) *La ville émergente*, La Tour d'Aigues, l'Aube éditions.

Dupuy G. (1991) *L'urbanisme des réseaux – Théories et méthodes*, Paris, Armand Colin.

Dupuy G. (1999) *La dépendance automobile*, Paris: Anthropos.

Dupuy G. and Bost F. (eds) (2000) *L'automobile et son monde*, La Tour d'Aigues: L'Aube.

Ellul J. (1954) *La technique ou l'enjeu du siècle*, Paris: Armand Colin.

Erikson R. and Goldthorpe J. H. (1992) *The Constant Flux – A Study of Class Mobility in Industrial Societies*, Oxford: Clarendon Press.

European Community (1995) *Un réseau pour les citoyens*, Bruxelles.

Ferman B. (1996) *Challenging the Growth Machine. Neighborhood Politics in Chicago and Pittsburgh*, Kansas, University Press of Kansas.

Forsé M. (1999) 'Social Capital and Status attaintment in contemporary France', in: *The Tocqueville Review-La revue Tocqueville*, n° 1 Vol.XX.

Fortier A.-M. (2000) *Coming Home: intersections of queer memories and diasporic spaces*, conference at Lancaster University.

Fouchier, V. (1997) *Les densités urbaines et le développement durable*, Paris: Editions du SGVN.

Friedmann J. (1995) 'Where we stand: a decade of world city research', in: Knox P. L. and Taylor P. J. (eds) *World Cities in a World System*, Cambridge: Cambridge University Press, 21-47.

Froud, J. et al. (2000) 'Les dépenses de motorisation comme facteur d'accentuation des inéga-lités et comme frein au développement des entreprises automobilles: une comparaison franco-anglaise', in: Dupuy, G. et Bost, F. (dir.) *L'automobile et son monde*, La Tour d'Aigues: Editions de l'Aube.

Fuétigné C. (1999) *Sociologie de l'exclusion*, Paris: L'Harmattan.

Gasparini G. (1995) 'On waiting', in: *Time and Society*, Vol. 4(1), 29-45.

Girard A. (1964) *Le choix du conjoint*, Paris: PUF.

Grafmeyer Y. *Sociologie urbaine*, Paris: Nathan.

Grafmeyer Y. et Dansereau F. (dir.) (1998) *Trajectoires familiales et espaces de vie en milieu urbain*, Lyon, Presses Universitaires de Lyon.

Harvey D. (1990) *The Condition of Postmodernity*, Oxford: Blackwell.

Hevan F. (2000) *Transports en milieu urbain: les effets externes négligés*, Paris: La Documentation Française.

Hochshild A. R. (1997) *The time bind. When work becomes home and home becomes work*, Metropolitan Books.

Jenks M., Burton E. and Williams K. (eds.) (1996) *The Compact City – A Sustainable Urban Form?*, London, E & FN SPON.

Jousselin, B. (1998) 'La mobilité résidentielle des ménages en 1994', in: Segaud M. et al. (dir.) *Logement et habitat, l'état des savoirs*, Editions de la découverte / textes à l'appuis, Paris, 120-127.

Joye D. Huissoud Th. and Schuler M. (eds.) (1995) *Habitants des quartiers, citoyens de la ville?*, Zurich: Seismo.

Judge D., Stocker G. and Wolman H. (eds.) (1995) *Theories of urban politics*, Sage: London, Thousand Oaks, New Dehli.

Jurczyk K. (1998) 'Time in women's everyday lives – between self-determination and conflicting demands', in: *Time and Society*, Vol.7(2), 283-308.

Kaplan C. (1996) *Questions of travel*, Durham and London: Duke University Press.

Karsenty B. (1994) *Marcel Mauss – Le fait social total*, Paris: PUF.

Kaufmann V. (1999) 'Mobilité et vie quotidienne: synthèse et questions de recherche', in: *2001 plus – Synthèses et recherches*, n° 48, Centre de Prospective et de Veille Scientifique, Direction de la recherche et des affaires scientifiques et techniques, Ministère de l'Equipement, des Transports et du Logement.

Kaufmann V. (2000) *Mobilité quotidienne et dynamiques urbaines – La question du report modal*, Lausanne: Presses polytechniques et universitaires romandes.

Kaufmann V. et Guidez J.-M. (1998) *Les citadins face à l'automobilité*, Lyon, Dossier du CERTU n° 80, CERTU.

Kaufmann V., Jemelin C. et Guidez J.-M. (2001) *Automobile et modes de vie urbains: quel degré de liberté?*, Paris: La Documentation Française.

Kaufmann V., Jemelin C. et Joye D. (2000) *Entre rupture et activités: vivre les lieux du transport* PNR41-A4, Zurich: EDMZ.

Keeling D. (1995) 'Transport and the World City Paradigm', in: *World Cities in a World System*, Cambridge, Cambridge University Press, 115-131.

Knox P.L. and Taylor P.J. (eds) (1995) *World Cities in a World System*, Cambridge: Cambridge University Press.

Kontuly, T. and Vogelsang, R. (1989) 'Federal Republic of Germany: the intensification of the migration turnaround', in: Champion A. (ed.) *Counterurbanization*, London, Edward Arnold, 141-161.

Kriesi H. (1994) *Les démocraties occidentales. Une approche comparée*, Paris, Economica.

Kudera W. and Voss G. (eds) (2000) *Lebensführung und Gesellschaft*, Opladen, Leske und Budrich.

Lahire B. (1998) *L'homme pluriel – les ressorts de l'action*, Paris: Nathan.

Latouche S. (1998) *Les dangers du marché planétaire*, Paris: Presse de Science.

Law J. and Hassard J. (eds.) (1999) *Actor Network Theory and After*, Oxford: Blackwell.

Lefèvre, C. et Offner, J.-M. (1990) *Les transports urbains en question – Usages – décisions – territoires*, Paris, éditions Celce.

Lemieux V. (1999) *Les réseaux d'acteurs sociaux*, Paris: PUF.

Lévy J. (1999) *Le tournant géographique*, Paris: Armand Colin.

Lévy J. (2000) 'Les nouveaux espaces de la mobilité', in: Bonnet M. and Desjeux D. (eds.) *Les territoires de la mobilité*, Paris: PUF, 155-170.

Lévy P. (1999) 'La pensée crash de Paul Virilio', in: *Les cahiers de médiologie*, Tribune du 1 mai.

Merton R. (1968) *Social theory and social structure*, 3rd ed., New York, Free press.

Mol A. and Law J. (1999) *Situated Bodies and Distributed Selves: Enacting Hypoglycaemia* published by the Department of Sociology, Lancaster University at: www.lancaster.ac.uk/sociology/stslaw5.html

Montulet B. (1998) *Les enjeux spatio-temporels du social – mobilité*, Paris: L'Harmattan.

Newman P. et Thornley A. (1996) *Urban Planning in Europe*, London: Routledge.

Offner J.-M. (1993) 'Les 'effets structurants' du transport: mythe politique, mystification scientifique', in: *L'espace géographique*, n° 3.

Offner J.-M. (2000) *Territorial Deregulation: Local Authorities at Risk from Technical Networks*, in: International Journal of Urban and Regional Research, Vol.24/1, 165-181.

Offner, J.-M. et Pumain, D. (dir.) (1996) *Réseaux et territoires – significations croisées*, La Tour d'Aigues: l'Aube éditions.

Ostrow J. (1990) *Social sensitivity: a study of habit and experience*, Albany: State University of New York Press.

Parsons T. (1952) *The Social System*, London, Routledge.

Parsons T. (1960) *Structure and Process in Modern Societies*, New York: Free press.

Pharaoh T. et Apel D. (1995) *Transport Concepts in European Cities*, Aldershot, Avebury.

Pooley C. and Turnbull J. (1998) *Migration and modility in Britain since the 18th Century*, London: UCL Press.

Potier F. (1996) *Le tourisme urbain: les pratiques des Français*, Arcueil: INRETS.

Prato P. and Trivero G. (1985) 'The spectacle of travel', The Australian Journal of Cultural Studies, 3.

Pucher, J. (1998) 'Urban transport in Germany: providing feasible alternatives to the car', in: *Transport Reviews*, Vol.18/4, 285-310.

Putnam R. (2000) *Bowling Alone*, New York: Simon & Schuster.

Rallet A. (2000) 'Communication à distance: au delà des mythes', in: *Sciences Humaines*, n° 104, 26-30.

Rapoport D. (1996) Interview in: Bellanger F. et Marzloff B. (eds.) *Transit*, La Tour d'Aigues, éditions de l'Aube: 271-275.

Rapport N. and Dawson A. (1998) 'The topic and the book', in: Rapport N. and Dawson A. (eds) *Migrants of Identity*, Oxford: Berg, 3-17.

Remy J. (2000) 'Métropolisation et diffusion de l'urbain: les ambiguïtés de la mobilité', in: Bonnet M. and Desjeux D. (eds.) *Les territoires de la mobilité*, Paris: PUF, 171-188.

Remy J. et Voyé L. (1992) *La ville: vers une nouvelle définition*, Paris: L'Harmattan.

Roch M. (1998) 'La spatialisation du social à l'épreuve de la mobilité: l'exemple de l'espace péri-urbain', in: *Espaces et Société* n° 94.

Salomon I., Bovy P. et Orteuil J.-P. (eds.) (1993) *A billion trips a day*, Kluwer, Dordrecht.

Sassen S. (1991) *The Global City*, Princeton: Princeton University Press.

Sayer A. (2000) *Realism and Social Sciences*, London: Sage.

Schuler M., Lepori B., Kaufmann V. and Joye D. (1997) *Eine Integrative Sicht des Mobilität – Im Hinblick auf ein neues Paradigma des Mobilitätsforschung*, Bern: Schweizerischer Wissenschaftsrat.

Schuler, M. et Joye, D. (1988) *Le système des communes suisses*, Office fédéral de la statistique, Berne.

Sheller M. and Urry J. (2000) 'The City and the Car', in: International Journal of Urban and Regional Research, Vol.24/4, 737-757.

Shields R. (1992) (ed.) *Lifestyle Shopping – the Subject of Consumption*, London, New York: Routledge.

Stalkev P. (2000) *Workers without frontiers*, London: Lynne Rienner.

Stimson R. J. and Minnery J. (1998) 'Why people move to the 'Sun-belt': a case study of long-distance migration to the Gold Coast Australia', in: *Urban Studies*, Vol.35 n° 2, 193-214.

Stoker G. (1995) 'Regime Theory and Urban Politics', in: Judge D., Stoker G. et Wolman H. (eds.) *Theories of Urban Politics*, London/Thousand Oaks/New Dehli, Sage Publications, 54-71.

Tarrius A. (1989) 'Perspectives phénoménologiques dans l'étude de la mobilité', in: Barjonet P.-A., 5 ed. *Transports et sciences sociales – questions de méthodes*, Caen: Paradigme, 47-81.

Tarrius A. (2000) *Les nouveau cosmopolitismes*, La Tour d'Aigues: L'Aube.

Tarrius A. (2001) 'Nouveaux territoires, nouveau cosmopolitismes', in: Bassand M. et al. (eds.) *Les enjeux de la sociologie urbaine*, Lausanne: Presses Polytechniques et Universitaires Romandes.

Tickamyer A. R. (2000) 'Space Matters! spatial inequality in future sociology', in: *Contemporary sociology*, Vol.29/6, 805-813.

Töpfler A. (1980) *The Third Wave*, New York: William Morow.

Touraine A. (1992) 'Inégalités de la société industrielle, exclusion du marché', in: Affichard J. and Foucauld J.-B. de (eds) *Justice sociale et inégalités*, Paris: Esprit.

Touraine A. (1995) *Critique of Modernity*, Oxford: Blackwell.

Urry J. (1990) *The Tourist Gaze*, London: Sage.

Urry J. (2000a) *Sociology beyond Societies, Mobilities for the Twenty First Century*, London: Routledge.

Urry J. (2000b) *Mobility and proximity, paper presented at the Mobilities group*, Dept. of Sociology, Lancaster University.

Veltz P. (1996) *Mondialisation, ville et territoires*, Paris: Presses universitaires de France.

Viard J. (1995) *La société d'archipel*, La Tour d'Aigues: l'Aube éditions.

Virilio P. (1984) *L'espace critique*, Paris: Christian Bourgois.

Virilio P. (1989) *Esthétique de la disposition*, Paris: Galilée.

Weber M. (1979) *Economy and society*, Vol. 1, Berkeley, University of California Press.

Wegener, M. and Fürst, F. (1999) *Integration of Transport and Land Use Planning*, Deliverable D2a – Land use transport interaction: state of the art. European Commission, 4ᵉ programme cadre de recherche, Bruxelles.

Wellman B. and Richardson R. (1987) *Analyse des réseaux sociaux – principes, développements, productions*, séminaire du CESOL, un niveau intermédiaire, les réseaux sociaux, Paris.

Whitelegg J. (1997) *Critical Mass*, London: Pluto.

Wiel, M. (1999) *La transition urbaine*, Mardaga, Sprimont.

Wiel M. and Rollier Y. (1993) 'La pérégrination au sein de l'agglomération brestoise', in: Les Annales de la recherche urbaine, n° 59-60, Paris, 151-162.

Wright E. O. & al. (1992) *The American Class Structure*, American Sociological Review, 6/47, 709-726.

Zarifian P. (1999) *Temps et modernité*, Paris: L'Harmattan.

Ziegler B. (1995) 'Zurich, ville modèle des transports urbains en Europe: la recette suisse', in: *Transport public*.

Index